SUSTAINABILITY COMMUNICATION ACROSS ASIA

Sustainability Communication across Asia distils the core components of environmental communication in the diverse milieu of Asian nations such as Indonesia, Malaysia, the Philippines, Thailand, and China.

The chapters in this book engage readers in a clear-sighted view of issues, challenges, and strategies related to sustainability communication in Asia, examining fundamental principles, digital strategies, and the role of language, as well as community engagement. The first part of the book features underpinning ideologies of sustainability communication. The authors go on to explore the prevalent trends and approaches in sustainable communication in the digital realm, examining the internet in general, social media, and gaming platforms. Finally, the book discusses the green efforts adopted among selected Asian communities, the role of communication, and the resulting societal impacts. Readers will be introduced to many related examples of Asian sustainability cases and issues that may differ from Western experiences. Interesting topics such as environmental gamification, edutainment and sustainability communication, and social media and sustainability are among those presented and elaborated at length by 21 writers with industrial and academic backgrounds.

Practical and inspiring, this book will be of great interest to students and scholars of sustainability and environmental communication, and Asian studies in general.

Mohamad Saifudin Mohamad Saleh holds a PhD degree in Environmental Communication from the Faculty of Sustainability, Leuphana University of Lüneburg, Germany. Upon graduation in September 2016, he started his career as a senior lecturer at the School of Communication, Universiti Sains Malaysia (USM). He was appointed the deputy dean of Academic, Career, and International at the same school in 2022. His areas of specialisation are environmental communication, sustainability communication, and media and environmental non-governmental organisations (ENGOs). He recently completed his research on Islamic environmental communication in Malaysia and sustainable university branding. He is currently leading national research on sustainable lifestyles in Malaysia.

Nur Atikah A Rahman obtained her PhD in Mass Communication from Universiti Putra Malaysia. Her education background includes a Master of Arts in Communication (Advertising Strategy) and a Bachelor of Communication (Persuasive Advertising) from Universiti Sains Malaysia. In June 2020, she attained the Foundation Certificate from the Institute of Practitioners in Advertising (IPA), Incorporated by Royal Charter, UK, which was funded by the Association of Accredited Advertising Agents Malaysia (4As). The focus of her current research is on self-branding in the social media realm, the role of social media influencers in brand loyalty, creative content strategies, and advertising.

Shaidatul Akma Adi Kasuma completed her PhD in Arts Education at the Centre of Education Studies, University of Warwick, UK. Prior to that, she obtained her Master's in Linguistics from Universiti Malaya, and Bachelor's in Science with Education (TESL) from Universiti Teknologi Malaysia. She is interested in the areas of technology in language learning, higher education, discourse analysis, and applied linguistics. Her publications are mainly on language learning education, and she looks forward to expanding her language and linguistics interest to more niche areas including language and sustainability communication in social media.

Routledge Studies in Environmental Communication and Media

Communicating Endangered Species
Extinction, News and Public Policy
Edited by Eric Freedman, Sara Shipley Hiles and David B. Sachsman

Communicating Climate Change
Making Environmental Messaging Accessible
Edited by Juita-Elena (Wie) Yusuf and Burton St. John III

Competing Discourses on Japan's Nuclear Power
Pronuclear versus Antinuclear Activism
Etsuko Kinefuchi

Essential Concepts of Environmental Communication
An A–Z Guide
Pat Brereton

Extreme Sustainability Rhetoric and Sustainable Development
Mauro Fracarolli Nunes and Camila Lee Park

Sustainability Communication across Asia
Fundamental Principles, Digital Strategies and Community Engagement
Edited by Mohamad Saifudin Mohamad Saleh, Nur Atikah A Rahman and Shaidatul Akma Adi Kasuma

For more information about this series, please visit: www.routledge.com/Routledge-Studies-in-Environmental-Communication-and-Media/book-series/RSECM

SUSTAINABILITY COMMUNICATION ACROSS ASIA

Fundamental Principles, Digital Strategies and Community Engagement

Edited by Mohamad Saifudin Mohamad Saleh, Nur Atikah A Rahman and Shaidatul Akma Adi Kasuma

Routledge
Taylor & Francis Group
LONDON AND NEW YORK

earthscan
from Routledge

First published 2023
by Routledge
4 Park Square, Milton Park, Abingdon, Oxon OX14 4RN

and by Routledge
605 Third Avenue, New York, NY 10158

Routledge is an imprint of the Taylor & Francis Group, an informa business

British Library Cataloguing-in-Publication Data
A catalogue record for this book is available from the British Library

ISBN: 978-1-032-31418-1 (hbk)
ISBN: 978-1-032-31419-8 (pbk)
ISBN: 978-1-003-30964-2 (ebk)

DOI: 10.4324/9781003309642

Typeset in Bembo
by Deanta Global Publishing Services, Chennai, India

CONTENTS

CONTRIBUTORS

Editors

Mohamad Saifudin Mohamad Saleh holds a PhD degree in Environmental Communication from the Faculty of Sustainability, Leuphana University of Lüneburg, Germany. Upon graduation in September 2016, he started his career as a senior lecturer at the School of Communication, Universiti Sains Malaysia (USM). He was appointed as the deputy dean of Academic, Career, and International at the same school in 2022. His areas of specialisation are environmental communication, sustainability communication, and media and environmental non-governmental organisations (ENGOs). He recently completed his research on Islamic environmental communication in Malaysia and sustainable university branding. He is currently leading national research on sustainable lifestyles in Malaysia.

Nur Atikah A Rahman obtained her PhD in Mass Communication from Universiti Putra Malaysia. Her education background includes a Master of Arts in Communication (Advertising Strategy) and a Bachelor of Communication (Persuasive Advertising) from Universiti Sains Malaysia. In June 2020, she attained the Foundation Certificate from the Institute of Practitioners in Advertising (IPA), Incorporated by Royal Charter, UK, which was funded by the Association of Accredited Advertising Agents Malaysia. Professionally, she previously worked as a copywriter for several advertising and branding agencies in the private sector before pursuing her career in academia. The focus of her current research is on self-branding in the social media realm, the role of social media influencers in brand loyalty, creative content strategies, and advertising.

Shaidatul Akma Adi Kasuma completed her PhD in Arts Education at the Centre of Education Studies, University of Warwick, United Kingdom. Prior to that, she obtained her Master's in Linguistics from Universiti Malaya, and Bachelor's in

Science with Education (TESL) from Universiti Teknologi Malaysia. She is interested in the areas of technology in language learning, higher education, discourse analysis, and applied linguistics. Her publications are mainly in language learning education, and she looks forward to expanding her language and linguistics interest to more niche areas including language and sustainability communication in social media.

Authors

Adnan Hussein is a professor in communication, Universiti Sains Malaysia, Penang. He completed his PhD in Communication at the University of Utah and Master's in Journalism and Mass Communication at Iowa State University. His interests and publications relate to big data, mass communication, media, and journalism studies. He was a vice rector for student affairs at Universiti Sains Malaysia. Prior to that, he was the dean of the School of Communication, Universiti Sains Malaysia. He is knowledgeable about the fields of media, journalism, and environmental communication. Therefore, he is often invited as a resource person to give insight and foresight on these across Indonesia, Thailand, and several other countries.

Anna Agustina is a senior lecturer at the Faculty of Communication, Universitas Pancasila. She completed her PhD in Environmental Communication at Universiti Sains Malaysia, and Master's in Journalism and Strategic Communication Management (2003) and Bachelor of Arts in Media and Journalism (1996), both from Universitas Indonesia. Her fields of interest are journalism, media studies, and environmental communication. Her work as a journalist, public relations officer, and communication specialist for several international programmes, such as Electricity and Restructuring Activities and Sustainable Environmental Assessment for Indonesian Spatial Policy, among others, ignited her interest in her fields of study. She is currently a coordinator of the Communication and Sustainable Development, chair of the UNESCO ORBICOM Network, Jakarta Chapter, Indonesia, and Dean of the Faculty of Communication, Universitas Pancasila.

Daryl Lustracion Superio is a registered professional librarian and is currently the Officer-in-Charge of the Library and Data Banking Services Section at the Aquaculture Department, Southeast Asian Fisheries Development Centre (SEAFDEC/AQD), Iloilo, Philippines. He does research on communication and media, and library and information studies. He has published several research papers on information-seeking behaviour, disaster management in libraries, bibliometrics, citizen science, information literacy, public libraries, and grey literature. He is an active board member of the Aquatic Sciences and Fisheries Abstracts (ASFA) and a member of the International Association of Aquatic and Marine Science Libraries and Information Centres (IAMSLIC). He gained his Bachelor's and Master's degrees in Library and Information Science from the Central Philippine University, Iloilo City, Philippines.

Debbita Tan Ai Lin completed her PhD in English Language Studies at Universiti Sains Malaysia (USM). She is a senior lecturer in the School of Languages, Literacies and Translation, USM. Prior to academia, she was a feature writer, with work published in various local and international publications. She is keen on interdisciplinary research, and her areas of interest include language learning, translation studies, and discourse analysis, with an emphasis on media and political discourse.

Eli Jamilah Mihardja obtained her PhD in Environmental Communication from Universiti Sains Malaysia in 2017. She is presently a senior lecturer on the Communication Studies Programme at Bakrie University and is also the head of the Geopark Study Centre at Bakrie University. Her research interests are primarily in the fields of place branding and environmental communication.

Emilia Bassar has been working as a communication and public relations expert, practitioner, and lecturer for more than 20 years. Since 2011, she has been involved in various work on climate change communication. These inspired her to initiate two regular events, the Climate Communication Forum (CCF) and Climate Change Communication Action (Aku Iklim) to facilitate knowledge, experiences, and best practices sharing among relevant stakeholders. She completed her Bachelor's in Public Relations at Prof. Dr. Moestopo University (Beragama), her Master's in Communication Management at the University of Indonesia, and PhD in Cultural and Media Studies at Gadjah Mada University. She was certified as an Indonesia-accredited public relations officer by the Public Relations Association of Indonesian in 2017 and 2020, and certified by the Climate Reality Leadership Corps from the Climate Reality Project in 2016 and 2021. In 2018, she founded the Centre for Public Relations, Outreach and Communication, which organises training, research, and consultancy in the field of communication and climate change.

Huang Miao is an associate professor at the School of Animation and Digital Arts, Communication University of China, Nanjing. He is a PhD student in the School of Communication, Universiti Sains Malaysia, under the supervision of Dr Mohamad Saifudin Mohamad Saleh and Dr Izzal Asnira Zolkepli. His main research areas are social media and game development.

Izzal Asnira Zolkepli is a senior lecturer in persuasive communication at the School of Communication, Universiti Sains Malaysia. She earned her PhD in Business Studies from the University of Malaya, Malaysia. Izzal's research is cross-disciplinary, at the intersection of digital marketing communication, digital consumer behaviour, and communication technology management. The focus of her research involves the convergence and adoption of communication technologies and proliferation of social media, mobile apps communication, and digital networked media, examining consumer engagement and technology acceptance. Owing to her active involvement in research and publication, she has recently had conferred on her the prestigious high-impact journal publication award known as Malaysia's Research Star Award by SCOPUS, in recognition of her outstanding performance and contribution to the research field in Malaysia.

Jamilah Ahmad is currently attached to the School of Communication, Universiti Sains Malaysia. She has produced over a hundred journal articles, proceedings articles, research monographs, books, and book chapters on a broad range of topics in the field of corporate social responsibility, environmental communication, public relations, youth studies, new media technology, and audience analysis. Contributing to the academic community, she has been actively sharing her expertise on editorial boards and as an issue reviewer for 48 international and local journals. Jamilah is also known for her extensive involvement as the principal investigator and co-researcher in over 50 research projects.

Joesyl Marie V. de la Cruz-Aranas is a special departmental coordinator and development communication staff of the Southeast Asian Fisheries Development Center/Aquaculture Department. She finished her Master's in Mass Communication (MMComm) and her Bachelor of Science in Development Communication at West Visayas State University's College of Communication in 2021 and 2011.

Mahadevan Krishnan is currently a lecturer in the School of Communication, Universiti Sains Malaysia in Penang, Malaysia. His areas of interest encompass public relations, corporate communication, sustainability communication, strategic corporate social responsibility, and organisational communication. With an interest in establishing links between business and sustainable initiatives, Mahadevan's research area encompasses the triple bottom line approach for strategic business positioning, which comprises sustainability in general, green initiatives, corporate imaging, identity and reputation, environmental sociology, and well-being communication.

Nurzali Ismail is a faculty member at the School of Communication, Universiti Sains Malaysia (USM). His areas of research include youth usage of digital media, sports communication, and crime. Prior to his current appointment, Nurzali was a visiting researcher at the Wee Kim Wee School of Communication and Information, Nanyang Technological University (NTU), Singapore. He was the recipient of an Australia Postgraduate Award (APA) and Monash Graduate Scholarship (MGS).

Nurul Selen Azizah ASP is an associate planner at Bappenas (Indonesian Ministry of National Development Planning). She has a Bachelor's degree in Urban and Regional Planning and an interest in spatial planning and social urban analysis. Currently, she is a member of the Geopark Action Plan Secretariat, Bappenas.

Rittipol Kuntasuwun obtained his PhD in Communication Arts and Innovation from the National Institute of Development Administration (NIDA), Thailand. His educational background includes a Master of Science in Marketing Management (International Programme) from the College of Management, Mahidol University, and a Master of Arts in English for Careers from the Thammasat University, Thailand. He also possesses a First-Class Honours Bachelor's degree in Communication Arts in Advertising from Bangkok University, Thailand. He has previously worked in corporate communication and stakeholders' management at PTT Public Company Limited, a Thai national oil and gas business. He is currently working at the executive level in marketing communications, branding management, sales and promotion, digital marketing, and data science analysis at PTT Oil and Retail Business Public Company Limited in Thailand.

Rona Dhel Cabrias Alingasa is an assistant professor and media practitioner turned educator. She is currently a faculty member at West Visayas State University in Iloilo City, Philippines. Her education background includes a Master's in Development Communication from UP Open University and a Bachelor of Arts in Broadcast Communication – Management from the University of the Philippines, Visayas. She is in the final stages of obtaining her PhD in Education at the University of San Agustin. Her research interests include media and information literacy, educational communication and technology, and community organising. She advocates for stakeholders' participation in development, as well as facilitating positive social change through appropriate information, education, and communication.

Suriati Saad is a faculty member in the School of Communication, Universiti Sains Malaysia. She graduated from Universiti Sains Malaysia with a PhD and a Master of Arts, and has a Bachelor's degree from Deakin University, Melbourne, Australia. Her research interests include corporate social responsibility, marketing communication, public relations, and media studies. Prior to joining the university, Suriati previously worked in the industry for 7 years as a senior executive at multinational companies such as Citibank, Motorola, and Escatec. She was a recipient of the Best Young Academician Award, from Emerald Group Publishing Limited, UK, awarded at the 11th International Conference on Corporate Social Responsibility, Lahti, Finland, in 2012.

Togu S. Pardede is a deputy director for geology, mining, and geothermal at Bappenas (Indonesian Ministry of National Development Planning). He has a PhD in Land Use Planning and Post Disaster Planning from the University of Tokyo. Now, he is a key person in the Indonesian Geopark National Action Plan.

Zhang Tong is working as a deputy manager for Nanjing Easyskill Technology Co., Ltd., China. She is also a lecturer at the School of Animation and Digital Arts, Communication University of China, Nanjing. Her main fields of interest are social media and digital marketing research.

INTRODUCTION

The 21st century places sustainability at the forefront of global planning and policy. In its widest definition, sustainability is the effort we make to meet our needs without compromising the ability of future generations to meet theirs in relation to the triad of economic, social, and environmental resources.

Universal emphasis on sustainability led to the 2015 development of the holistic Sustainable Development Goals (SDGs) by the United Nations General Assembly. The SDGs underline 17 agendas: SDG 1 (No Poverty), SDG 2 (Zero Hunger), SDG 3 (Good Health and Well-being), SDG 4 (Quality Education), SDG 5 (Gender Equality), SDG 6 (Clean Water and Sanitation), SDG 7 (Affordable and Clean Energy), SDG 8 (Decent Work and Economic Growth), SDG 9 (Industry, Innovation and Infrastructure), SDG 10 (Reduced Inequality), SDG 11 (Sustainable Cities and Communities), SDG 12 (Responsible Consumption and Production), SDG 13 (Climate Action), SDG 14 (Life below Water), SDG 15 (Life on Land), SDG 16 (Peace, Justice, and Strong Institutions), and SDG 17 (Partnerships for the Goals).

Sustainability is not, therefore, a novel concept and has been practised in the Western hemisphere to ensure the well-being of the current and future generations. Westerners prioritise sustainable lifestyles in their daily conduct and take sustainability matters into account in their developments, planning, and policies. Research and education on sustainability work hand in hand to cultivate sustainable citizens as human capital. The same cannot be claimed for many developing countries. Despite the global pervasiveness and intentions of the SDGs, the concepts of sustainability and sustainable development are still not in sync with their society's lifestyles and behaviour, especially among the middle class and the underprivileged. While corporations and higher education institutions might be aware of the sustainable concept, sustainability research and discourses are fundamentally rhetorical and exclusive to academics and scientists. Indeed, there is a growing uneasiness over the depleting resources, economic downturn, and deprived society among citizens across many nations; however, the lack of awareness and understanding of the notion of sustainability frequently results in fleeting changes and temporary motivation. For example, many Malaysians relate sustainability to only environmental matters, without recognising that it involves economic and societal discourses too.

An underdeveloped, imperfect understanding of sustainability results in unsustainable practices and hampers sustainability communication efforts. Sustainability communication is the commitment made by an organisation to disseminate information about its sustainable goals, practices, and strategies to its stakeholders. Sustainability communication is often circulated via internal and external means: annual reports, press releases, websites, and social media posts. However, this information will be meaningless to those unfamiliar with the concept of sustainability and an organisation's sustainable goals.

Like sustainability, the concept of communicating sustainability intention also derives from Western perspectives. In Asian countries, the practice of sustainability communication is new and foreign to the mass public, scientists, and academics alike, causing a gradual progression compared with the West. The lack of formal references and resources on sustainability communication across Asian countries contributes to the lower public literacy and interest in such matters. Consequently, though some communities and organisations do practise sustainability endeavours across the triads of environment, society, and economy, the lack of awareness of such activities as sustainable behaviours leads to short-term, one-off actions, instead of continuous, impactful ones. As sustainability endeavours will only be successful when the messages are received by the audience, the process of sustainability communication is of the utmost importance in raising the public's awareness of ecological living.

Scholars debate the relevance of sustainability communication in Asian countries owing to the limited enforcement of sustainability policies and lifestyles among their citizens. This, in fact, is a misguided view, as education and continuous interaction are key to transforming global citizens into a more mindful and conscious society. While there are many sustainable communities and organisations across the Asian regions, their ecological conduct may not be visible owing to the lack of communication. In an effort to showcase Asian countries' sustainability communication efforts, this book addresses the issues, trends, and practices in selected nations: Malaysia, Indonesia, Thailand, the Philippines, and China. Notably, all the authors are scholars affiliated with universities and industries that advocate sustainable practices and sustainable communication, and they have written their chapters based on their observation, experience, reflection, and case studies of their immediate environment.

This book exposes global and Asian readers, mainly scholars and students, to the present and future of sustainability communication in Asia. It is novel and significant, highlighting the Asian perspective to provoke universal discussions on the issue. Scholars and students who want to learn about sustainability communication in Asia can use the book as their reference, especially for academic purposes. With a good book that focuses on sustainability communication in the Asian context, students and scholars, not only from Asia, will be able to utilise the book for discussion. This will balance the view of sustainability communication between the West and Asia.

The book features quality articles centring on three themes – principles of sustainability communication, digitisation and integration of sustainability communication, and community engagement for sustainability – all of which are discussed in the immediate Asian milieu.

The first part of the book focuses on the ideologies underpinning sustainability communication, before moving on to unlock the prevalent trends in and approaches to sustainable communication in the digital sphere through the internet, social media, and gaming platforms. Lastly, the book narrates the green efforts adopted by selected Asian communities and their impact on society. Practical and inspiring, the book will support readers on their journey in sustainability communication efforts, whether in research or their personal and professional lives.

This book is organised into 14 exciting chapters with three themes. Theme 1, principles of sustainability communication, comprises five chapters. Chapter 1 conceptualises sustainability communication and narrates the critical differences between sustainability communication in the West and Asia. Chapter 2 highlights the language of sustainability communication and how words impact the public's perceptions of an organisation. Chapter 3 introduces the development of a sharing economy and its roles in promoting sustainability communication and practices in China. Chapter 4 demonstrates the strategic sustainability communication model of companies listed on the Stock Exchange of Thailand. Chapter 5 discusses the importance of persuasion techniques in sustainability communication.

Theme 2, the digitisation and integration of sustainability communication, underlines the following. Chapter 6 focuses on sustainability communication, consumers, and social media. Chapter 7 deliberates on the practices of Malaysian gaming influencers in sustainability communication. Chapter 8 explores the role of a geopark virtual exhibition in sustainability communication in Indonesia. Chapter 9 discusses the essentials for sustainable mobile applications in China.

Finally, Theme 3, community engagement for sustainability, encompasses these topics. Chapter 10 highlights the factors that encourage journalists to embark on sustainability communication. Chapter 11 talks about the role of community radio in imparting climate change information to its public. Chapter 12 focuses on sustainability from the perspectives of corporate social responsibility. Chapter 13 highlights the edutainment concept and its association with sustainability communication. Lastly, Chapter 14 describes the sustainability information desires and search behaviour among fishers in the coastal communities in the Philippines.

It is hoped that the wide-ranging topics presented in this book about sustainability communication across Asia will accentuate the efforts of involved entities to make the world a better place and pave the way for more sustainability discourses in the near future. In balancing sustainability discourses from the Western regions, this book is an innovative attempt that gives voices to Asian writers and a glimpse of sustainability communication practices in Asian organisations. This book will be beneficial for those who are interested in delving further into the realms of sustainability communication and its development from a non-Western perspective.

1

SUSTAINABILITY COMMUNICATION

West vs Asia

Mohamad Saifudin Mohamad Saleh

Introduction: Understanding Sustainability Communication

In brief, "sustain" is a term that is derived from the Latin word *sustinere*, which means hold or maintain. The term "sustainability" was coined by the famous German scientist Hans Carl von Carlowitz (1713), in his seminal work on "*nachhaltige waldwirtschaft*" (sustainable forest economy). In it, he emphasised the importance of the "*nachhaltige entwicklung*" (sustainable development) principle for forest management (Wilderer, 2007, p. 2).

Universally, sustainable development has been defined by the Brundtland Report of the World Commission on Environment and Development as a concept that "meets the needs of the present without compromising the ability of future generations to meet their own needs" (World Commission on Environment and Development, 1987). The primary goal of sustainability is therefore to ensure the well-being of present and future generations (Tapia-Fonllem, Corral-Verdugo & Fraijo-Sing, 2017). In line with the idea of sustainable development, sustainability is frequently referred to as an effort to strike a balance between three elements – environmental, social, and economic (Boyer et al., 2016). Discussions on sustainability in the mid 1980s were heavily focused on the environment and economy (Partidario, Vicente & Belchior, 2010), but have since evolved to include social issues as factors that may also affect the well-being of the environment and economy.

Following the general term, sustainability, which is largely Western-centric, sustainability communication is a term that represents the essence of human interaction with the natural and social environment (Godemann & Michelsen, 2011). In the most straightforward understanding, sustainability communication refers to communication on both social and environmental issues (Lodhia, 2014), and also economic. In sustainability communication, people hold sustainability

DOI: 10.4324/9781003309642-1

discourse through print or digital platforms, such as the media, to improve the quality of life for both present and future generations (Barchilon, 2011). As a scholarly field, sustainability communication mainly focuses on strategic communication or marketing, and, thus, public communication usually involves the media's role in delivering sustainability information to the public (Weder, Lemke & Tungarat, 2019). By conducting sustainability communication, changes in sustainability, either small or significant, are to be expected, with short- or long-term effects (Janoušková et al., 2019).

There are two distinctive types of sustainability communication, which concern communication of, and communication about, sustainability. The former relates to sender-oriented sustainability messages from one sender to many receivers, while the latter is discourse-oriented and focuses on debates and discussions on sustainability messages (Fischer et al., 2016). An example of sustainability communication on corporate sustainability is a report written by a company for its stakeholders (Fischer et al., 2016). In contrast, an example of communication about sustainability is an editor's letter published on media platforms, or interactions on television that promote discussions on the topic (Newig et al., 2013).

The elaboration above clearly shows the substantial conceptualisation and discussion of sustainability communication by Western scholars. However, sustainability communication is considered a relatively new and sometimes overlooked concept among Asians – especially in developing countries. As an attempt to delineate sustainability communication, Malaysian scholars such as Mohamad Saleh (2016) define the concept as sending sustainability messages or information from the perspectives of environmental, economic and social, or societal issues from the sender to a receiver. This definition, however, follows the basic Western-centric communication model that has been proposed by Shannon (1948), also known as the model of communication process, whereby communication (including sustainability communication) is transmitted from the sender to receiver. Therefore, it is not surprising that sustainability communication has not been featured as part of the essential communication skills module in higher education institutes in Malaysia (Sivapalan et al., 2017). However, it is safe to conclude that the conceptualisation of sustainability communication is similar in the West and in Asia, as sustainability is a universal issue that concerns communities around the globe.

Integrative Approach to Sustainability Communication

Sustainability communication originates from environmental communication as it expands the concept of the environment from sustainable development (Fischer et al., 2016). This results in an overlap of the two communication areas. However, as discussed above, sustainability communication should focus not only on the environment, but also on socioeconomic issues such as health, poverty, and gender, among others, in line with the 2030 Sustainable Development Goals

(SDGs). Environmental communication is a process of communicating environmental messages via various channels that encourage public debate (Zikargae, 2018). By employing environmental communication, people can obtain, comprehend, and utilise environmental messages (Bakar, 2015) to increase public awareness about the environment and risks associated with it (Wu et al., 2021). Typically, the specific environmental issue that is communicated is based on the location and the country. For example, the haze phenomenon that affects large parts of Malaysia and some parts of Indonesia is often communicated by the media or environmental non-governmental organisations in these countries. This issue may rarely be reported by the West as its environmental communication may instead centre on discourses concerning global warming and climate change.

Besides environmental communication, it is crucial to also note that sustainability communication is also integrated with climate change communication. This is because climate change is one of the main focuses under the SDG 13, climate action. Like the previous two communication areas, climate change communication stresses the importance of dialogue and engagement between the communicator and the public to spur climate action (Lucas, 2021). In other words, it cannot rely on communication of sustainability (sender-oriented), but can rely more on communication about sustainability (discourse-oriented). Although climate change communication is globally prominent, in Malaysia, it is considered more a seasonal communication that is usually spread by the media during well-publicised events such as international climate change conferences. Owing to this, South East Asian nations are deemed to be under-informed about the complexities and urgency of climate change (Chainan & Sutthisima, 2021).

Sustainability issues and problems are also interrelated with the natural sciences (Kruse, 2011), making sustainability communication a multidisciplinary area that, besides environmental and risk communication, is also connected with science communication (Weder, Lemke & Tungarat, 2019). Science communication is a process of communicating scientific issues to the public. In line with environmental and climate change communication, science communication also emphasises dialogue about science between the communicator and recipient to increase the public's science literacy. Similar to environmental communication, science communication is integrated with climate change, though it can also be positioned within scientific, environmental, or even sustainability discourses depending on the angle of the discussion. However, some scientific issues, such as medicine-related issues, are specifically meant for science communication.

The discussion on the integrative approach of sustainability communication towards environmental, climate change, and science communication brings us to the conclusion that the West and Asia delineate the concept in the same way. This is indeed a very straightforward concept, but must be addressed so that the similarities and differences between the areas are clear.

Online Sustainability Communication

Like many other types of communication, sustainability communication has also shifted from traditional to online platforms. This shift was even more pronounced during the period of the COVID-19 pandemic, as global citizens were confined in the comfort of their homes, using the internet and smartphones to search and share information, including that related to sustainability. The fast and vast development of online platforms makes it hard to differentiate between the online sustainability communication of Western and Asian countries. In other words, the existence of online sustainability communication itself has removed the boundaries between countries, including those separating Asia and the West. Nevertheless, online sustainability communication can differ from one country to another, particularly in its approach and medium of use (Santos, Rodrigues & Branco, 2016). For instance, countries such as China have blocked their citizens from accessing certain social media websites such as Facebook and Twitter, thus preventing sustainability communication messages from being disseminated through these platforms.

The online website is one of the platforms used widely for communicating sustainability messages. Websites are useful for sustainability communication as they are interactive platforms that allow for effective and fast delivery (Ghanem & Elgammal, 2017). Firms, for instance, have mostly used websites to communicate sustainability messages to their stakeholders, especially through sustainability reporting. Lodhia (2014), in his research on three Australian mining companies, discovered that cost-efficiency (economy factor) is one of the factors that influenced the companies to use a website for their sustainability communication activities to engage stakeholders. In terms of a comparison between the Western and Asian contexts, past researchers have also reported on the difference between sustainability communication via websites by South Korean and United States of America (USA) firms. Specifically, South Korean firms displayed more power distance features in their sustainability communication than their American counterparts (Ki & Shin, 2015).

Notwithstanding, nowadays, there are many social media sites available, such as Facebook, Twitter, and TikTok, that serve as powerful tools for online sustainability communication. The nature of real-time communication via social media has led to easier dissemination of accessible and reachable sustainability messages to global audiences without boundaries. For example, sustainability messages from the West can reach audiences in other parts of the world within seconds or minutes. This is important as, in certain countries, especially in Asian regions, the governments have exerted control over sensitive sustainability issues on traditional media platforms. Alternatively, social media allow the public to freely communicate and share information, including on sustainability or green messages that may be deemed delicate or controversial.

However, in communicating sustainability information via the online platforms of social media, the sender needs to have the creativity and knowledge to attract the attention of other users, especially the youth. The sustainability messages on social media need to be concise and attractive, so that the audience is discouraged from skipping or skimming over them. For example, in their study of eco-friendly hotels and sustainability communication via social media, Kapoor, Balaji, and Jiang (2021) suggested that hotels could use sensual appeal messages such as "It feels good to conserve. Make a green choice" for their social media posts to induce feelings of positivity and hope among the travellers who opt for a sustainable holiday experience. On the other hand, a lack of appeal in sustainability messages will discourage audience interest in the messages.

Likewise, images such as pictures, graphs, diagrams, and sketches, can always be utilised by professional communicators and organisations in their online sustainability communication, as visuals are more powerful at attracting attention compared with text. For instance, instead of typing lengthy descriptions of dirty rivers, the communicator can use a picture of a polluted river. The visual will help the audience to quickly understand the severity of the matter and invoke feelings that could potentially induce the desired action. Xue and Muralidharan (2015), for instance, found that the use of green visuals in advertising could generate an ecological image in audiences, therefore steering them to associate the brand with eco-friendliness characteristics.

Culture as an Enabler for Sustainability Communication

There are many aspects that can influence the process and success of sustainability communication. One of the biggest influences is the culture of the community, especially within the context of sustainability communication in diverse Asian contexts, as they are concerned with inherent values and beliefs. It is understandable that culture cannot be separated from sustainability (communication) as the achievement of a sustainability goal depends on human action that is influenced by the culture itself (Soini & Dessein, 2016). In other words, the cultural aspect has a significant impact on sustainable development (Nocca, 2017) and, therefore, sustainability communication.

Unlike the more homogeneous societies in the West, many Asian countries are heterogeneous in their demographic makeup. For instance, the culture in countries such as Malaysia and Singapore is shaped by Malay, Chinese, and Indian cultures, among others, that inevitably influence the public's understanding of, and approach towards, sustainability communication.

In Malaysia, for example, Mohamad Saleh et al. (2021) found that newspapers such as *Berita Harian* and *Utusan Malaysia* instilled Islamic values in their environmental reporting to suit the majority of Malay readers who are Muslims. This was done by citing verses from the Quran and hadith (deeds and sayings of the Prophet Muhammad) that are linked with the reported environmental issues. Injecting such religious elements is a way to make Muslim readers feel more connected with the news than would be achieved by a piece of conventionally written environmental news.

In addition, understanding the culture in one specific country is crucial as it will determine the effectiveness of sustainability communication to the local population. In China, for example, the idea of sustainability and sustainable development is not new as it is embedded in the traditional Chinese culture via the concept of harmonious development (Li et al., 2016). Hence, it is critical for the sustainability communicator to understand and draw upon the local cultural beliefs and values in communicating sustainability matters to increase effectiveness and heighten engagement.

Likewise, there are many unique communities in Asia, including the culturally rich, nature-centric indigenous people. For example, in Taiwan, the indigenous Smangus tribe has its own term or traditional thought called "Gaga" that represents the way of life, including the tribe's norms, rituals, knowledge, and penalties that link it to nature and exhibit its relationship with the land (Yu, 2018). Without understanding this unique community and its unique culture, sustainability communication cannot be done successfully. In fact, the risk of promoting a sensitive message while communicating sustainability issues could occur if we do not carefully understand the local culture and community.

The Role of Universities in Sustainability Communication

Education on sustainability communication at the tertiary level is important to produce informed and adept sustainability communicators among university students. By enrolling in sustainability communication courses either at the undergraduate or postgraduate level, students would be exposed to fundamental knowledge and critical skills to effectively communicate complex and abstract sustainability messages to the public. More importantly, exposure to sustainability communication at university or college during the developmental phase can also help cultivate necessary character traits and values such as good attitudes, beliefs, and behaviours to support sustainability (Lertpratchya et al., 2017).

A number of higher institutions in the West offer sustainability communication courses or degrees at the university. For instance, Jönköping University in Sweden has its own Master's degree specifically for sustainable communication that focuses on the three dimensions of sustainability – environment, social, and economy. The course aims to produce experts in the field to obtain the Sustainable Development Goals. In Canada, Royal Roads University offers its undergraduate students a sustainability communication course. On the course, students are exposed to principles and approaches for disseminating sustainability messages via different platforms. Similarly, the University of Tasmania, Australia, offers a course called Communicating Sustainability that aims to nurture students' knowledge and skills in communicating sustainability to stakeholders and audiences.

Besides sustainability communication, some universities offer their students environmental communication courses. The Swedish University of Agricultural Sciences in Sweden, for example, offers a Master's degree in environmental communication and management that educates students in effective communication and management of the environment and sustainability; it also emphasises the importance of collaboration between different actors in the community. Interestingly, Yale University in the USA has its own dedicated centre for environmental communication and related research, known as the Yale Center for Environmental Communication. A similar research centre can be found at the Leuphana University of Lüneburg, Germany, where the Institute of Environmental and Sustainability Communication (INFU) has been operating for more than 15 years.

It is clear that Western society possesses many environmental and sustainability communication experts who have actively assisted in the promotion of sustainability communication programmes, whereas none of the universities in Asian countries offers such packages. This is largely owing to the fact that the fields of sustainability and sustainability communication are still new and developing in Asia. The limited number of Asian scholars with expertise in sustainability communication further contributes to the reduced prominence and popularity of the field.

Notwithstanding, there is an exception in the Universiti Sains Malaysia (USM), located in Penang, Malaysia, which offers a course called Communication for Sustainable Development. This course is compulsory for students majoring in Communication Studies in the School of Communication and may be taken as an minor subject for students from other schools/faculties at USM. The course complies with USM's sustainable and green university agenda that promotes sustainability knowledge among the students. Similar to its Western counterparts, the Communication for Sustainable Development course aims to educate students on the importance of effective communication of sustainability messages to the public, as well as the numerous approaches that can be used to achieve this. The course not only presents sustainability themes occurring in Malaysia, but also covers pertinent discussions in the wider contexts of Asia and the world.

As sustainability communication targets sustainable behavioural transformation, universities in Asia play crucial roles in promoting auxiliary education and activities related to it. In other words, sustainability communication should be institutionalised and perceived as an important subject by Asian universities. At present, sustainability and sustainable development as subjects are emphasised by many universities across Asia, though they have yet to be framed from the communication perspective. Thus, universities in Asia should consider offering more sustainability communication-related courses to address this growing need.

Sustainability Communication as a Legitimacy Instrument for Organisations

The literature defines legitimacy as a form of reflection of the relationship between an organisation and its surroundings (Patel, Xavier & Broom, 2005). Legitimacy is also a sign that indicates an organisation's capacity to achieve its goals (Yoon, 2005). Besides that, legitimacy is defined as a perception, like desirability and appropriateness, formed by the public about an organisation (Aerts & Cormier, 2009) – in other words, a kind of public approval of the organisation's existence in the society (Herlin, 2015). On a larger scale, legitimacy reflects an organisation's reputation (Deephouse & Carterand, 2005) and is, therefore, crucial for its long-term survival (Mousa & Hassan, 2015).

Nowadays, regardless of profit, organisations strive to maintain legitimacy (Wodahl, Ogle & Heck, 2011). To accomplish this, organisations must be able to satisfy their stakeholders' expectations by becoming more accountable and transparent (Bednárová, Klimko & Rievajová, 2019) and, most importantly, align their activities with societal values (Mousa & Hassan, 2015), social norms, and laws and regulations (Tang, 2017). Hence, as long as an organisation adheres to the laws, is transparent, and is beneficial to the society, it will gain the stakeholders' trust and be considered legitimate (Panwar, Hansen & Kozak, 2014).

Sustainability communication is a crucial tool for an organisation to maintain its legitimacy in the public eye. Conway, O'Keefe, and Hrasky (2015), in their study of Australian NGOs, discovered that the organisations managed to maintain their legitimacy via annual reports about tsunami relief efforts. In Asia, the local authorities in Indonesia, for instance, have disclosed their integrity information to the public via their website in an effort to show visibility and gain legitimacy (Joseph et al., 2019). In Malaysia, news organisations' reporting on environmental issues has also made them be publicly perceived as credible and ecological (Mohamad Saleh, 2016).

By the same token, it is important to note that illegitimacy can occur owing to undesirable organisational behaviour (Patterson & Allen, 1997). For example, an oil company risks facing perceptions of illegitimacy in the event of involvement in an oil-spill incident, which can result in a tarnished reputation (Summerhays & De Villiers, 2012). In this case, a legitimacy gap is created when an organisation's actions are not in line with societal expectations (Deegan, 2006). A legitimacy crisis can occur when certain non-profit organisations establish partnerships with commercial companies that divert the former from their mission, which consequently leads them to lose their stakeholders' support (Herlin, 2015). An example of such a case was reported by Liu, Zhang, and Yao (2021) in China, where the authors discovered that tourists would cancel their travel plans to a destination with a previous record of unethical activities. Therefore, an organisation must reduce or even close the legitimacy gap or crisis by assessing the kinds of behaviours that are considered appropriate by its society, to ensure continual enhancement and improvement of its legitimacy (Panwar, Hansen & Kozak, 2014). For instance, in responding to a diesel scandal, Volkswagen avoided panic and re-established its moral legitimacy by evoking desired emotions in its audience (Florio & Sproviero, 2020). Based on this premise, it can be reasonably concluded that, in both Western and Asian contexts, sustainability communication, especially via environmental messages, can be a good instrument for an organisation to create and gain legitimacy. On the other hand, improper sustainability communication can jeopardise an organisation's reputation, trust, and image in the eyes of its stakeholders.

The Barriers to Sustainability Communication

Sustainability as a concept is considered highly complex and ambiguous (Newig et al., 2013). To comprehensively embrace it, an individual is required to engage in debates involving technology, society, and politics (Partidario, Vicente & Belchior, 2010). The sustainability communicator, therefore, needs to be creative, specifically in using appealing messages and images (Strother & Fazal, 2011). For instance, celebrity or social media influencers who support green or sustainable agendas can be featured in sustainability communication messages in both traditional and social media. Leonardo DiCaprio is a prominent example of this. As a celebrity who advocates on climate change issues, he makes the message more appealing and attractive to both Western and non-Western audiences.

Furthermore, within the Asian context, the challenge of sustainability communication also concerns a perceived lack of interest in related issues, as topics on the economy and politics are usually given the spotlight. As many developing Asian countries still struggle with basic necessities and facilities for growth and survival, sustainability and sustainable developments are not their top priority, with the exception of specialised groups of academics, researchers, and nation leaders. This is a challenge that is perhaps unique to Asian countries, as the developed countries in the Western hemisphere have more visible and established sustainability lifestyles and the Sustainable Development Goals that are familiar to the public.

Across Asia, the lack of media attention to sustainability issues poses yet another challenge. In Thailand, for instance, media coverage on environmental themes is low compared with reports on other news such as politics, entertainment, economics, and sports (Chainan & Sutthisima, 2021). Similarly, in Malaysia, the media only give attention to environmental or sustainability issues in the event of an environmental disaster or a public event, such as the Paris Agreement. The media are mostly interested in politics, entertainment, economy, and sports, as these are considered more newsworthy. The lack of media attention to sustainability issues, therefore, presents an obstacle to effective sustainability communication endeavours being developed in Asian countries.

Likewise, the public's attention to, interest in, and knowledge of sustainability matters are also pertinent to effective sustainability communication. Insufficient knowledge, especially among residents of rural areas, has, however, hampered efforts to communicate sustainability messages owing to the lack of comprehension. In Bangladesh, Kabir et al. (2016) studied 6720 households from rural communities and discovered that nearly half of the respondents (45.8%) did not have any knowledge of climate change and claimed to have never heard about the concept, despite the country's vulnerability to the crisis. Similarly, in

Yogyakarta, Indonesia, Sulistyawati, Mulasari, and Sukesi (2018) found in their study that there was a low and inconsistent understanding of climate change and its effects on health among 508 adolescents. These findings reveal the real challenges in communicating sustainability messages to communities who may be completely unaware that such issues exist, let alone have heard of them.

Finally, another barrier worth discussing is related to misinformation or the widespread dissemination of fake news on sustainability on social media. Misinformation or fake news is a serious phenomenon that occurs globally, and sustainability communication is not exempt. Misinformation, also known as false information, continues to be a prominent feature in the communication about COVID-19, to the extent that it is labelled as being as dangerous as the virus itself (Rastogi & Bansal, 2021). From the perspective of sustainability communication, especially communicating about SDG 3, good health and well-being, fake news on COVID-19, particularly about the vaccines, creates confusion and induces fear and negative feelings among the public. Consequently, many global citizens still refuse to acknowledge the existence of the virus, strongly reject vaccination, and denigrate its effectiveness. In Pakistan, for instance, misinformation on COVID-19 is spread widely by religious leaders, politicians, experts, and even lay users on social media such as Twitter, Facebook, WhatsApp, and TikTok (Ittefaq, Hussain & Fatima, 2020).

Without a doubt, the characteristics of social media, such as the availability of the share button feature, have removed the role of a gatekeeper, which consequently facilitates the widespread dissemination of misinformation on sustainability. Thus, users need to be educated on the literacy and ethics of using social media, especially on the importance of fact-checking before they attempt to share any information. At the same time, experts such as scientists, academics, and researchers need to be trained to effectively frame their sustainability messaging to increase the level of credibility. Basch, Yalamanchili, and Fera (2022) found widespread misinformation on climate change on TikTok and suggested the importance of credible professionals to counter the fake news.

Conclusion

In summary, the discussion on sustainability communication is primarily drawn from the Western point of view, and this chapter debates the fundamental comparisons of sustainability communication in the West with that in Asia. Overall, it is fair to infer that the discussion of sustainability communication focusing on Asian contexts is somewhat limited, as the area is new and emerging, and may have yet to yield substantial empirical and conceptual literature.

Sustainability communication in Asia and in the West is quite similar, particularly in terms of its conceptualisation. Despite the universality of the concept, sustainability communication may manifest in distinct ways in Asian countries owing to the versatility and dynamic cultures and values compared with the more secular Western approaches. Continuing the comparison, the role that Asian universities play in sustainability communication is still relatively highly limited. Asian higher education settings must give serious consideration to courses, programmes, or packages related to sustainability communication for pushing a sustainability agenda that leads to transformed, ecological societies. This is indeed a long process, which is why this chapter calls for more experts and researchers working in Asian contexts to engage in debates surrounding sustainability communication.

References

Aerts, W., & Cormier, D. (2009). Media legitimacy and corporate environmental communication. *Accounting, Organizations and Society, 34*, 1–27.

Bakar, D. A. (2015). Good, the bad and the ugly: Framing the country development through environmental communication. *Procedia-Social and Behavioral Sciences, 168*, 8–12.

Barchilon, M. G. (2011, October 17–19). Sustainability communication tracks in technical communication: Our value in deriving sound, sustainable solutions. In 2011 IEEE International Professional Communication Conference. IEEE.

Basch, C. H., Yalamanchili, B., & Fera, J. (2022). #Climate change on TikTok: A content analysis of videos. *Journal of Community Health, 47*, 163–167.

Bednárová, M., Klimko, R., & Rievajová, E. (2019). From environmental reporting to environmental performance. *Sustainability, 11*, 1–12.

Boyer, R., Peterson, N., Arora, P., & Caldwell, K. (2016). Five approaches to social sustainability and an integrated way forward. *Sustainability, 8*, 1–18.

Chainan, P., & Sutthisima, S. (2021). Climate change communication in Southeast Asia from journalist perspective in the context of Thailand: Raising awareness by environment communication. In C. Yamahata, D. M. Seekins, & M. Takeda (eds). *Social transformations in India, Myanmar, and Thailand: Volume I.* Singapore: Palgrave Macmillan.

Conway, S. L., O'Keefe, P. A., & Hrasky, S. L. (2015). Legitimacy, accountability and impression management in NGOs: The Indian Ocean tsunami. *Accounting, Auditing and Accountability, 28*(7), 1075–1098.

Deegan, C. (2006). Legitimacy theory. In Z. Hoque. *Methodological issues in accounting research: Theories and methods.* London: Spiramus Press.

Deephouse, D. L., & Carterand, S. M. (2005). An examination of differences between organizational legitimacy and organizational reputation. *Journal of Management Studies, 42*(2), 329–360.

Fischer, D., Lüdecke, G., Godemann, J., Michelsen, G., Newig, J., Rieckmann, M., & Schulz, D. (2016). Sustainability communication. In H. Heinrichs, A. Wiek, P. Martens, & G. Michelsen (eds). *Sustainability science. An introduction.* Netherlands: Springer.

Florio, C., & Sproviero, A. F. (2020). Repairing legitimacy through discourses: Insights from the Volkswagen's 2015 diesel scandal. *Meditari Accountancy Research*, 1–19.

Ghanem, M., & Elgammal, I. (2017) Communicating sustainability through a destination's website: A checklist to inform, motivate, and engage stakeholders. *Journal of Travel & Tourism Marketing, 34*(6), 793–805.

Godemann, J., & Michelsen, G. (2011). Sustainability communication - an introduction. *Sustainability communication: Interdisciplinary perspectives and theoretical foundations.* Berlin: Springer.

Herlin, H. (2015). Better safe than sorry: Nonprofit organisational legitimacy and cross sector partnerships. *Business & Society, 54*(6), 822–858.

Ittefaq, M., Hussain, S. A., & Fatima, M. (2020). COVID-19 and social-politics of medical misinformation on social media in Pakistan. *Media Asia, 47*(1–2), 75–80.

Janoušková, S., Hák, T., Ne˘cas, V., & Moldan, B. (2019). Sustainable development - a poorly communicated concept by mass media. Another challenge for SDGs? *Sustainability, 11*, 1–20.

Joseph, C., Gunawan, J., Madi, N., Janggu, T., Rahmat, M., & Mohamed, N. (2019). Realising sustainable development goals via online integrity framework disclosure: Evidence from Malaysian and Indonesian local authorities. *Journal of Cleaner Production, 215*, 112–122.

Kabir, Rahman, M. B., Smith, W., Lusha, M. A. F., Azim, S., & Milton, A. H. (2016). Knowledge and perception about climate change and human health: Findings from a baseline survey among vulnerable communities in Bangladesh. *BMC Public Health, 16*, 1–10.

Kapoor, P. S., Balaji, M. S., and Jiang, Y. (2021). Effectiveness of sustainability communication on social media: Role of message appeal and message source. *International Journal of Contemporary Hospitality Managementc* 949–972.

Ki, E-J., & Shin, S. (2015). Organization sustainability communication (OSC): Similarities and differences of OSC messages in the United States and South Korea. *Computers in Human Behavior, 48*, 36–43.

Kruse, L. (2011). Psychological aspects of sustainability communication. In J. Godemann & G. Michelsen (eds). *Sustainability communication: Interdisciplinary perspectives and theoretical foundations*. Berlin: Springer.

Lertpratchya, A. P., Besley, J. C., Zwickle, A., Takahashi, B., & Whitley, C. T. (2017). Assessing the role of college as a sustainability communication channel. *International Journal of Sustainability in Higher Education, 18*(7), 1060–1075.

Li, Y., Cheng, H., Beeton, R. J. S., Sigler, T., & Halog, A. (2016). Sustainability from a Chinese cultural perspective: The implications of harmonious development in environmental management. *Environment, Development and Sustainability, 18*, 679–696.

Liu, Y., Zhang, R., Yao, Y. (2021). How tourist power in social media affects tourism market regulation after unethical incidents: Evidence from China. *Annals of Tourism Research, 91*, 1–13.

Lodhia, S. (2014). Factors influencing the use of the world wide web for sustainability communication: An Australian mining perspective. *Journal of Cleaner Production, 84*, 142–154.

Lucas, C. H. (2021). Climate friction: How climate change communication produces resistance to concern. *Geographical Research, 60*(3), 1–12.

Mohamad Saleh, M. S. (2016). *Environmental communication among the media and environmental non-governmental organisations (ENGOs) in Malaysia: A roller coaster ride?* PhD Dissertation, Leuphana University of Lüneburg, Germany.

Mohamad Saleh, M. S., Adi Kasuma, S. A., Zuknik, M. H., & Nik Hasan, N. N. (2021). Islamic values in environmental communication within the Malaysian media: A case study of Utusan Malaysia and Berita Harian. *Jurnal Komunikasi: Malaysian Journal of Communication 37*(3), 54–72.

Mousa, G. A., & Hassan, N. T. (2015). Legitimacy theory and environmental practices: Short notes. *International Journal of Business and Statistical Analysis, 2*(1), 41–53.

Newig, J., Schulz, D., Fischer, D., Hetze, K., Laws, N., Lüdecke, G., & Rieckmann, M. (2013). Communication regarding sustainability: Conceptual perspectives and exploration of societal subsystems. *Sustainability, 5*, 2976–2990.

Nocca, F. (2017). The role of cultural heritage in sustainable development: Multidimensional indicators as decision-making tool. *Sustainability, 9*, 1–28.

Panwar, R., Hansen, E., & Kozak, R. (2014). Evaluating social and environmental issues by integrating the legitimacy gap with expectational gaps: An Empirical assessment of the forest industry. *Business and Society, 53*(6), 853–875.

Partidario, M. R., Vicente, G., & Belchior, C. (2010). Can new perspectives on sustainability drive lifestyles? *Sustainability, 2*, 2849–2872.

Patel, A. M., Xavier, R. J. & Broom, G. (2005). Toward a model of organizational legitimacy in public relations theory and practice. Paper presented at International Communication Association Conference, 26-30 May 2005, New York.

Patterson, J. D., & Allen, M. W. (1997). Accounting for your actions: How stakeholders respond to the strategic communication of environmental activist organisations. *Journal of Applied Communication Research, 25*, 293–316.

Rastogi S., & Bansal D. (2021). Misinformation analysis during COVID-19 pandemic. In M. Tuba, S. Akashe, & A. Joshi (eds). *ICT systems and sustainability. Advances in intelligent systems and computing, Vol 1270.* Singapore: Springer.

Santos, S., Rodrigues, L. L., & Branco, M. C. (2016). Online sustainability communication practices of European seaports. *Journal of Cleaner Production, 112*, 2935–2942.

Shannon, C. E. (1948). A Mathematical Theory of Communication. *Bell System Technical Journal, 27*(3), 379–423.

Sivapalan, S., Afsharian, A., Shafiq, N., Nuruddin, F., & Sivapalan, M. (2017, November). Communicating sustainability: Context matters. In 2017 7th World Engineering Education Forum (WEEF). IEEE.

Soini, K., & Dessein, J. (2016). Culture-sustainability relation: Towards a conceptual framework. *Sustainability, 8*, 1–12.

Strother, J. B., & Fazal, Z. (2011). Can green fatigue hamper sustainability communication efforts?. In 2011 IEEE International Professional Communication Conference, 1–6. IEEE.

Sulistyawati, S., Mulasari, S. A., & Sukesi, T. W. (2018). Assessment of knowledge regarding climate change and health among adolescents in Yogyakarta, Indonesia. *Journal of Environmental and Public Health*, 1–7.

Summerhays, K., & De Villiers, C. (2012). Oil company annual report disclosure responses to the 2010 Gulf of Mexico oil spill. *Journal of the Asia-Pacific Centre for Environmental Accountability, 18*(2), 103–130.

Tapia-Fonllem, C., Corral-Verdugo, V., & Fraijo-Sing, B. (2017). Sustainable behavior and quality of life. In G. Fleury-Bahi, E. Pol & O. Navarro. *Handbook of environmental psychology and quality of life research*. Switzerland: Springer Cham.

Tang, Y. (2017). A summary of studies on organizational legitimacy. *Open Journal of Business and Management, 5*(03), 487–500.

Weder, F., Lemke, S., & Tungarat, A. (2019). (Re)storying sustainability: The use of story cubes in narrative inquiries to understand individual perceptions of sustainability. *Sustainability*, 11, 1–15.

Wilderer, P. A. (2007). Sustainable water resource management: The science behind the scene. *Sustainability Science, 2*, 1–4.

Wodahl, E. J., Ogle, R., & Heck, C. (2011). Revocation trends: A threat to the legitimacy of community-based corrections. *The Prison Journal, 91*(2), 207–226.

World Commission on Environment and Development. (1987). *Our common future.* Oxford: Oxford University Press.

Wu, M., Long, R., Bai, Y., & Chen, H. (2021). Knowledge mapping analysis of international research on environmental communication using bibliometrics. *Journal of Environmental Management, 298*, 1–12.

Xue, F., & Muralidharan, S. (2015). A green picture is worth a thousand words?: Effects of visual and textual environmental appeals in advertising and the moderating role of product involvement. *Journal of Promotion Management, 21*(1), 82–106.

Yoon, Y. (2005). Legitimacy, public relations, and media access: Proposing and testing a media access model. *Communication Research, 32*(6), 762–793.

Yu, C.-Y. (2018). An application of sustainable development in indigenous people's revival: The history of an indigenous tribe's struggle in Taiwan. *Sustainability, 10*, 1–20.

Zikargae, M. H. (2018). Analysis of environmental communication and its implication for sustainable development in Ethiopia. *Science of the Total Environment, 634*, 1593–1600.

2

LANGUAGE AND SUSTAINABILITY COMMUNICATION

Shaidatul Akma Adi Kasuma

Introduction: Sustainability Issues and Communication in Malaysia

Sustainability communication is a business strategy for companies that have integrated sustainability into their operational and strategic activities. It gives a company a platform to inform its stakeholders about its business, operations, and management of a sustainable approach (Bhatia, 2020).

Sustainability communication is a fairly new concept in Asia, and particularly in Malaysia. It is more familiar among the big corporations and higher education institutions that participate in international discourses but foreign to many small, local businesses. The government, non-governmental organisations (NGOs), and educational institutions in Malaysia have actively collaborated in matters of sustainable development via a concerted effort and contextualised strategy to produce professional workforces to manage resources sustainably and efficiently (Foo, 2013). Some tertiary institutions that advocate sustainability initiatives include Universiti Sains Malaysia (USM), Monash University Malaysia, Universiti Malaya (UM), INTI University College (INTI-HC), Universiti Malaysia Pahang (UMP), and Universiti Putra Malaysia (UPM). Meanwhile, sustainable industry collaborators include Panasonic, GreenTech, Toyota, and Tesco, which has won an award for creating a future zero-carbon supermarket. One of its initiatives involves eco-friendly recycle bags to replace single-use plastic bags (Foo, 2013).

Issues of environmental sustainability became prominent in 2021–2022 when some parts of the nation were hit by severe floods and landslides owing to alleged massive deforestation and logging activities. Even more worrying, these anthropogenic factors cause other natural disasters, such as soil erosion and flash and

DOI: 10.4324/9781003309642-2

mud floods, that contribute to the loss of biodiversity and habitat (Abd Majid et al., 2021; Ibrahim, 2022).

Other environment-related issues in the Malaysian area are single-use plastic bags and straws. In 2019 and 2020, the government of Malaysia banned single-use drinking straws in Kuala Lumpur, Putrajaya, and Labuan, in support of the "No Plastic Bag" campaign. The government encourages consumers to choose biodegradable alternatives to reduce pollution and preserve human and animal lives. In both campaigns, the Malaysian university student population demonstrated a positive attitude towards the initiatives and indicated awareness of harmful plastic substances and threats (Lee & Chuan, 2012), but were still inclined to choose plastic drinking straws and bags, instead of reusable stainless steel straws, wooden materials, and biodegradable grocery bags (Abd Hamid & Wan Yahaya, 2020; Mohd Abdul Kadir et al., 2020). Genç (2017) termed this behaviour symbolical support of sustainability despite following non-sustainable agendas. Though informed of the threats, Malaysians do not seriously act on environmental concerns and are not willing to jeopardise their lifestyle and convenience.

As a result, Ogiemwonyi et al. (2019) concluded that Malaysian consumers had an unhealthy lifestyle and lack of attentiveness towards the environment, despite massive national green campaigns by the government and NGOs. From the tourism point of view, Tölkes (2020) concluded that consumers do not feel a responsibility to act to realise sustainability initiatives and avoid the opportunity, though aware of the benefits. This liberates them from the deactivation of a moral obligation to behave sustainably, while putting the blame on industry. In the case of single-use shopping bags, the imposed levy of MYR0.20–0.50 (US$0.06–0.12) significantly reduced plastic usage and encouraged consumers to bring their own bags and food containers. As such, Malaysians' green efforts largely hinge on money-related disadvantages. This substantiates Mefalopulos's (2005) and Tölkes's (2020) observation that economically challenged consumers and citizens may not be able to fully appreciate global ecological concerns when they face a daily struggle for survival owing to limited financial resources or behaviour specific-knowledge.

Apart from environmental issues, social and economic sustainability are a familiar discourse in Malaysia. The economic well-being and welfare of the B40 group is a paramount theme. B40 is defined as the bottom 40% of Malaysian household incomes, households who earn less than US$1103 per month. The B40 communities are prioritised in all aspects of economic and social life, including education, job and business opportunities, sustenance, household living aid, housing, electric rebate programmes, and transportation.

Accordingly, discourses on other sustainability issues are present but are rarely continuous or strongly championed by brands, organisations, or the mass media. For example, the issues of refugees and stateless children have been brought up now and again by NGOs and academics, but not picked up by businesses. A study by Rahmat et al. (2021) on the educational rights of stateless children in Malaysia is progressive, as it calls for the amendment of existing laws and policies, for example, the Zero Reject Policy, 2019, to reflect those of Thailand that allow stateless children to enrol in public schools without any prerequisites. An ecological policy is valuable and effective when it receives some degree of attention from the mass media, but the framing process is impacted by those who have power and access to the discourse (Genç, 2017). The media in Western countries have more freedom to report and criticise government policies or corporations' decisions on environmental, societal, and economic matters. The same cannot be said for non-Western countries, such as Malaysia, that restrict the media's and NGOs' freedom to research and write about many environmental issues, especially those linked to household names and controversial projects, such as the Lynas rare earth risks (Nik Hasan & Dauda, 2020).

Other varieties of sustainability, such as corporate sustainability in Malaysia's ICT sector, were addressed by Pranugrahaning, Donovan, Topple, and Masli (2020). They emphasised and reiterated the lack of empirical studies and paucity of research in the area. Likewise, earlier, Saadatian, Lim, Mat, and Sopian (2012) stated the limited studies that underlie Malaysia's efforts and initiatives in the realm of sustainability, despite its vision to transform into a developed country by 2020, following the Kyoto Protocol ratification on climate change. Saadatian et al. (2012) opined that Malaysia must fulfil the requirements across different aspects of sustainable development, besides the economy and technology.

Therefore, while the environmental aspect of sustainability seems to be executed impressively by conglomerates and more participative entities, there is a question as to whether their intention was communicated successfully. In the age of social media that encourage two-way communication and continuous dialogues, sustainability advocates are given various platforms to express their ideas, objectives, and actions. It is important that ecological discourse is contextualised to the local communities in building a resilient community. This is because, in sustainable discourses, sociocultural approaches are as important as political-economic approaches. Humans and the environment are central to any social-change activities, and they function within their immediate milieu. Therefore, an emphasis on local systems, norms, and culture that shift from solely Western-led development is crucial to developing a sustainable change (Servaes & Lie, 2015).

The Language of Sustainability Communication

Communicating sustainability intention is not easy, as it encompasses content knowledge articulated via conscientious linguistics selection and persuasive appeals. This often requires the expertise of professional communicators who are able to illustrate and verbalise an organisation's concept, imagination, and ideas (Byrum, 2019). In the words of Stibbe (2009), a sustainability communicator's task is to match words with the worldview, ideology, and passions of those who hear them.

Herrick and Pratt (2013) connected sustainability language to Austin's and Searle's illocutionary act in deriving the term "discourse". To them, discourse forms the fundamentals of sustainability communication as it can use many genres and cross-disciplines to impart meaning. In communicating ecological ideology, constitutive discourse or persuasive narrative appeals to human nature that is deeply rooted in, and influenced by, sociocultural values and variations (Genç, 2017; Herrick & Pratt, 2013). As an intrinsic human instinct, narrative enquiries are commonly employed in the field of media and communication (Bernier, 2020; Weder, Lemker, & Tungarat, 2019).

Narrative or storytelling, as a way of knowing and communicating, allows members of the public to immerse themselves in and empathise with an ecological situation (Weder et al., 2019). When the issue is fully understood, remedies may be offered for behavioural change. Stories convey and evoke emotions and explain problems, and, thus, Veland et al. (2018) saw narratives depicting transformed futures as able to structure and shape human comprehension and imagination. At the same time, they fill in the deficit for more serious sustainability issues. Fischer et al. (2020) coined the term "SusTelling" in defining "storytelling for sustainability". Effective narration of issues may sustain a desired behaviour or state of being and empower action by the audience.

Narratives have a high persuasive value. The degree of persuasion changes according to the message presented. As part of the illocutionary act, assertive commands are commonly found in environmental messages, in phrases such as "Stop illegal logging" and "Plant more trees". A message can be persuasively assertive and well received when the audience perceives it as important. However, should it lack conviction, a more suggestive appeal is required to reassure audience members (Kronrod et al., 2012). This is done by employing linguistic strategies, via words such as "natural" and "organic", new words with the prefix "eco-", and the use of the verb "to love" in association with the green market (Johannessen et al., 2010; Peirson-Smith & Evans, 2017). Nevertheless, this approach is more common in advertising than sustainability communication.

Problems related to narration, according to Peirson-Smith and Evans (2017) and Yeldar (2022), are not only that promotional sustainability messages often lack specific meaning and lead to confusion and uncertainty, but also that ecological discourse regularly adopts a negative narrative approach using metaphors to illustrate the catastrophic outcomes of careless consumption of resources. Consumers are then left to construct their own understanding, which may diverge from the actual message. In line with this, Yeldar (2022) lists eight clusters of words and phrases used repeatedly as crutches when companies communicate sustainability, resulting in unclear messages. These are: "our commitment", "the future", "the planet", "biggest challenge", "our journey", "together, we can", "building a better", and "goal for business". Educators and communicators face similar struggles in distinguishing conceptual ambiguity between storytelling and pedagogical approaches when sustainability themes are discussed (Fischer et al., 2020).

In these cases, the excessive self-promotion of oneself as eco-conscious (greenwashing), as well as association with green images, videos, and language (e.g. 100% recycled paper, 5-star energy rating), may backfire and be branded as false, misleading claims; hence, this should be avoided (Johannessen et al., 2010; Lyon & Montgomery, 2013). It is regarded as deceptive ecological persuasion and puts a firm at a disadvantage, resulting in loss of consumers and lower profits (Jha & Verma, 2022). Words such as "ethical" and "organic" evoke a cynical response from Hong Kong residents, indicating their uncertainty and disbelief, when they relate to the fashion world (Johannessen et al., 2010). Therefore, words with a broader meaning and more accessibility to the audience, such as "eco", "green", and "environment", should be included when promoting ecological discourse, though "environment" is often associated with adverse weather and ecological dysfunction. Nevertheless, it was reported that organisations that practised what they preached wrote in more complex styles, whereas those who dissociated actions from statements produced more sweeping statements (Crilly, Hansen, & Zollo, 2016). Accordingly, Newman, Pennebaker, Berry, and Richards (2003) found that deceptive communications are written using simpler structures than truthful communications.

Likewise, the term "sustainability" is too broad for singular interpretations, which leads to varying (in)action by consumers. Therefore, the word should be used with specific, lucid, contextual meaning in sustainability communication to provoke action. The opposite of greenwashing is greenhushing, whereby companies downplay sustainability messages and complex issues, as well as normalising sustainability, to reduce customer guilt. Greenhushing communication highlights customer benefits using explicit, affective, experiential, and active language to legitimise customers' hedonistic use of facilities (Font et al., 2017). Similarly, Stibbe (2009) argued against "sustainability correctness", which is a form of political correctness that dictates the "right" sustainability words and phrases to be used in mitigating the seriousness of the climate change situation. Instead, sustainability communicators should rethink adopting alternative discourse from a diverse range of existing sustainability literature that may or may not promote different worldviews.

The effectiveness of sustainability messages is also dependent on the messenger. Messages from an established entity or organisation are deemed more trustworthy and credible. For example, sensuous sustainability messages from eco-hotels evoke positive emotions of fun, pleasure, excitement, relaxation, happiness, optimism, and altruism compared with social media influencers' messages. In contrast, sustainable messages with guilt appeal written by social media influencers were better received by the audience, while the ones published by eco-hotels were considered to be preaching and exaggerating (Kapoor et al., 2021).

Message Construction in Communication

Sustainability message construction makes conscious effort to gain the audience's trust and confidence in an organisation's good intentions (Jha & Verma, 2022).

An effective sustainability message, like any other communication, combines content and language at the macro level (meaning, purpose, organisation, and cohesion of ideas) and micro level (grammar and vocabulary).

In his study of meaning mapping, Brown (2013) quoted Halliday's three-layer view of language – i.e., meaning (semantics), sound (phonology), and wording or lexicogrammar (syntax, morphology, and lexis). In written sustainability communications, lexicogrammar forms the inner core of a language and is put together to realise the stratum of meaning (semantics or discourse semantics). Austin (1976; cited in Nwagbara and Belal, 2019) theorised that grammar (micro level) performs at both the conscious (macro level) and the unconscious level and shapes organisational discourses or practices (meso level). In discourse semantics, meanings constructed via the lexicogrammar of a text are important in shaping the audience's mental representation (Brown, 2013; Crilly et al., 2016). In this regard, lexicogrammar and semantics are seen as forming a functional, meaning-making system (systemic functional linguistics, SFL) that looks into the relationship between language and its function in social settings, instead of being treated as isolated and discontinuous. The relationship is dependent on social and cultural context, where the language use is purposeful and by choice (social semiotics). In the words of Halliday (1977, p. 195), "By 'text' … we understand a continuous process of semantic choice. Text is meaning and meaning is choice".

Accordingly, the lexicogrammar and semantics exist within a certain register, defined as the characteristic of a certain domain of communication. A register formed by the macro (sociocultural) and micro (lexicogrammatical) aspects is realised via three functions, namely field, tenor, and mode. Field explains the nature of the social activity, such as the purpose and subject matter of text; tenor focuses on the relationship between participants; and mode highlights the channel of communication. Each of these registers is compatible with a linguistic metafunction – i.e., ideational (field), interpersonal (tenor), and textual (mode) (Martin & Rose, 2007).

Like SFL, critical discourse analysis (CDA) also views language production at the macro and micro levels as purposeful. The fundamentals of CDA dissect explicit and implicit discursive choices, in an attempt to legitimise social inequalities and power (Mullet, 2018). CDA is characterised as,

> a problem-oriented focus, an emphasis on language, the view that power relations are discursive, the belief that discourses are situated in contexts, the idea that expressions of language are never neutral, and an analysis process that is systematic, interpretive, descriptive, and explanatory.
>
> *(Mullet, 2018, p. 120)*

Accordingly, CDA is employed in the examination of many political texts, as well as government policies, to identify the fundamental ideology and inclination of the discourse.

Both semantics and lexicogrammar are vital in the construction of an effective sustainability communication, as congruous linguistics adaptation results in positive perceptions of a firm's ability and intention and heightens credibility. Distinct combinations of these aspects create different materials and genres – such as academic articles, essays, stakeholders' reports, advertorials, newsletters, brochures – that have specific functions. For example, stakeholders' reports inform and reassure stakeholders about a firm's sustainability initiatives, whereas social media communications persuade and engage mass audiences. Advertorials promote a firm's service or product, while accentuating its sustainability initiatives. The corporate annual report (CAR) has its own language structure and formality compared with adverts that appear in magazines (Ndlovu, 2020; Sayyed Hussain et al., 2020). What they may share is the positive framing of sustainability messages, using adverbs, action verbs, and adjectives, to highlight the rationality and benefit of the product/service, as well as to illustrate and verbalise effort.

Content Analysis in Sustainability Communication

Language undeniably shapes perceptions and reality. As ecological issues become more accessible via social media sharing, a community of green consumers arises and forces corporations to adapt to a more sustainable approach in their business activities. Corporations respond to these pressures by identifying as ecologically friendly, projecting a green image to consumers. Ecologically aligned companies are regarded as ethical, superior, refined, and more appealing to eco-conscious consumers. A rebranding of the British Petroleum company from a traditional oil company into a green energy service provider positively impacted the public's perceptions of it (Cox, 2008). In the tourism industry, sustainable achievements are highlighted on more sophisticated websites to indicate superiority (Tiago et al., 2021). Consequently, consumers are willing to pay more to secure quality organic or natural products (Valentine, 2020).

Sustainability keywords in tweets and posts published on social media have been examined to determine the frequency and effectiveness of a firm's sustainability communication efforts. These keywords included *Earth Day*, *environment protection*, *climate change*, *plastic-free*, *air pollution*, *biodiversity*, *endangered species*, *recycling*, *workplace safety*, and *renewable energy* (Jha & Verma, 2022). While these keywords may provide a point of convergence of sustainability communication across regions, there are, undoubtedly, distinctive sociocultural values practised by local communities and institutions, and other sustainability themes. For example, some Western and Asian companies similarly showcased their institutional corporate social responsibility (CSR) messages on corporation websites. However, the Asian firms were less inclined to engage with themes related to "organisation and management", "CSR interactivity", and "business care" or to ethically dissociate CSR effort and "business care" (Vollero et al., 2022). Another distinction between the Western and non-Western firms is the limited extent of environmental discussion (i.e. climate change) that was present on the non-Western organisations' corporate web pages owing to legal concerns and less media pressure (Fernández-Vázquez & Sancho-Rodríguez, 2020). The construction of an effective sustainability communication is, therefore, highly dependent on regional culture and norms. The content, language, and visuals to support an ideology must be coherent and cohesive to appeal to the public and align with their altruistic values and perceptions.

However, there are concerns over the legitimacy of execution: whether a company does more "talking" or is truly "walking" the initiatives. Unfounded claims often lead to a mismatch between a company's proclaimed identity and the image perceived by the public (Jha & Verma, 2022; Johnson & Greenwell, 2020; Saran & Shokouhyar, 2021; Valentine, 2020). In a text mining study, Saran and Shokouhyar (2021) conducted frequency analyses of sustainability words tweeted by automakers and the public to differentiate between the green corporate image (the public's perception) and identity (the corporations' brand) on Twitter. The words most frequently tweeted by corporations were "hybrid", "electrical", "performance", "fuel", and "environmental", and by the public, "electrical", "sustainability", "environmental", "energy", "hybrid", "performance", "emissions", "energy", and "fuel". Not only did the public tweet more green content than the corporations, there is a significant gap between the companies' claims about their green identity and how the public perceived such an image. Therefore, the corporations' approach to addressing green issues significantly differed from that of the public and, hence, highlighted the importance of professional content communicators in crafting and conveying the most accurate messages to warrant similar conceptualisations of a brand's image and identity. In contrast, a content analysis on Twitter showed that companies who truly performed sustainability initiatives in the UK were, in fact, underselling their efforts, although they possessed a higher climate leadership rating with no false sustainability claims (Johnson & Greenwell, 2020).

Linguistics Analysis in Sustainability Communication

In sustainability communication, a persuasion technique is key to coaxing the public's ecological transformation. Nwagbara and Belal (2019) combined communication studies' persuasive theory with CDA to examine the persuasive language in Nigerian oil companies' CSR reports at the micro, meso, and macro levels. A persuasive strategy at the single word level was found to be most effective in CSR reports, followed by a persuasive phrasal strategy and a persuasive chain of words strategy. At the single word level, nominalisation of sustainable development "relies on the process of naming things, places, policies, ideas, qualities, and events among others to launch rhetoric of persuasion and ideology" (p. 2407). Other single words that were found to persuasively construct the portrayal of "responsible organisations" among problematic Nigerian oil companies were verbs, adjectives, and adverbs, besides nouns and pronouns. This study highlights the importance of linguistic choice, as the fundamental of SFL, in the construction of an organisation's discourse to demonstrate legitimacy, accountability, and engagement, reflecting a specific social order; it also highlights the persuasive significance of a message that translates into the portrayal of a company's image.

Grammar choices are often intuitive, reducing deliberate manipulation of the way content is expressed (Hart, 2014; Ireland & Pennebaker, 2010). The lexicocognitive perspective adopted by Crilly et al. (2016) focuses on how speakers express themselves (lexicogrammar) rather than what they say. Grammatical aspects that influence the audience's mental representations include tones (positive or negative), adjectives, adverbs, tenses, pronouns, voices (active or passive), verbs, and nominalisations. For example, pronouns distinguish speakers from hearers, marking social identity (e.g. "we" vs "them"), and tenses indicate focus on time and probabilities (e.g. "had", "might"). The integration of these aspects in the framing of messages can also create exclusive (e.g. "versus", "but", and "only") and inclusive (e.g. "and" and "additionally") tones. To avoid sounding exclusive in their messages, organisations often exaggerate and make grand claims in their sustainability messages (Crilly et al., 2016).

The growing importance of social media in channelling sustainability messages increasingly blurs the lines between formal and informal forms of writing, where the final material of a CAR was inter-discursive, resembling a hybrid of a report and promotional materials (Sayyed Hussain et al., 2020). Oftentimes, poor performance or a negative overview is downplayed and mitigated by distancing self/product and putting the blame on external factors. Processing verbs, inanimate nominal groups, nominalisation, and passive voice are often utilised for this purpose (Ndlovu, 2020; Sayyed Hussain et al., 2020).

Visual Analysis in Sustainability Communication

Social media are rooted in multimodality. Sustainability communication on social media encourages the use of texts, supported by visuals in the forms of images and symbols to convey meaning. The combination of visuals and language is indeed attractive to social media users, as Huang et al. (2019) discovered that infographics with sustainability themes received the highest engagement and response rate on both Twitter and LinkedIn. In language studies, semiotics theory is commonly employed to examine the way humans use signs to construct meaning of the world around them. Images and symbols are treated as meaningful signs that are purposefully positioned to convey messages. Semiotics is multidisciplinary and extensively used in public service advertising (PSA) and visual design.

In relation to "greenwashing", Fernández-Vázquez (2021) used CDA, ecolinguistics, and multimodal analysis to determine how global companies that produced the highest carbon emissions enforced visual narratives and ideologies in presenting as concerned agents over climate change. They compared corporations' and NGOs' sustainability messages on websites and found that the former's use of sustainability images was more limited than the latter's. The images that appeared on the corporations' websites downplayed human agency and interpersonal functions and disregarded issues caused by economic development and industrial activities. Instead, the images emphasised the importance of technology in mitigating environmental crises. The image choice distanced and detached the corporations from human relations. Therefore, instead of relying on "greenwashing" strategies, the corporations suggested that they empower communities' involvement in the process of environmental preservation to develop a more favourable reputation and higher social standing.

Case Study: Universiti Sains Malaysia's Sustainability Communication on Facebook

In this small-scale case study, the author analysed the posts published on Universiti Sains Malaysia's Facebook page (SFL mode) for sustainability language and content, based on the words and phrases adapted from Jha and Verma's (2022) research. USM has one of the most active Facebook pages among the public universities in Malaysia to connect with existing and potential students, alumni, and other stakeholders (SFL tenor). USM monitors sustainability in various aspects of life and aims to boost the billions with the lowest socio-economic status via education and preserve the ecosystem via research.

There were 134 posts on Universiti Sains Malaysia's Facebook page over the span of 4.5 months (November 2021 to mid March 2022). The themes (SFL field) often published on the Facebook page are the university's activities in officiating in collaborations with other institutions, top management's visits, announcements of academic and social events, introductions to new facilities, USM's surroundings and campus ground, and celebrations of staff's and students' achievements. These posts were often accompanied by colourful photos and videos to support the message and spread information. The highest number of "likes" received by the page was 5900 for a post captioned, "Here we go again guys to the season. 'universiti SAKURA malaysia'", accompanied by six photos of trees with pink blossom in campus grounds that resembled Japan's *sakura*. The hashtags for this post were *#usm*, *#welead*, *#sakura*.

Out of the 134 entries, only five posts were found to contain the environmental sustainability-related words and phrases listed by Jha and Verma (2022). These words and phrases are; *sustainable*, *sustainability*, *green*, *recycling*, *saving energy*, *solar panel*, and *solid waste management*. They appeared across posts with various themes, which were Deans and Directors' Induction, Courtesy Visit to EU Ambassador, A New Learning Space at USM, Energy Saving as One of USM's Sustainable Efforts, and Strategic Corporation between USM and SWCorp in National Solid Waste Management.

When used, the sustainability-related words and themes mainly appeared as nouns, which makes the posts more objective and trustworthy. Only "green" was used as an adjective, which is a step in the right direction, as excessive usage of adjectives will make the posts sound like sales pitches and lower their credibility. While the posts conveyed good intentions, the low frequency of sustainable-related posts and detached environment-linked words resulted in monotonous messages. The word "sustainability" is too abstract and impersonal to motivate people into taking immediate action (Stibbe, 2009). Limited persuasive messages were observed, and the posts simply reported information instead of influencing the community to practise a green lifestyle. Therefore, the environmental sustainability theme was less impactful on USM's Facebook page.

Staying true to its identity as a leading higher education institution, USM posted frequent updates that centred on research initiatives, innovations, teaching and learning facilities, staff's and students' achievements, and industry collaborations. USM sets its vision as "Transforming Higher Education for a Sustainable Tomorrow" and its mission as "a pioneering, transdisciplinary research intensive university that empowers future talents and enables the bottom billions to transform their socio-economic well-being". In accordance with these, it accentuates more of the economic and social sustainability themes on its Facebook page. This is realised via economic- and social sustainability-related words and phrases like *student mobility/exchange programme, quality education, assisting underprivileged children's education, student entrepreneurship, Program Santuni Gelandangan* (Homeless Welfare Programme), *special needs children, cybersecurity, university–industry collaboration, volunteerism, to empower community, to balance the socioeconomic gap, bounce back from burnout,* and *prisoner rehabilitation.* The themes of economic and social sustainability are diverse, ranging from student entrepreneurship, quality teaching and learning, graduate employability, and physical and emotional well-being to industry and institutional links and collaborations, as well as the university's daily activities. In December 2021, the 58th USM convocation ceremony was held. For the whole month, the Facebook page was dedicated to announcing and celebrating graduates' achievements via daily updates of graduation photos and videos. This shows USM's efforts to celebrate students' mobility to improve quality of life.

From the perspective of message construction and lexicogrammatical choices, some posts possessed examples of persuasive values and inclusivity via narratives. An example of a post written on 10 November 2021 is shown below:

> Tablet Donation Ceremony to B40 Staff (Household) with children of 7–12 years old. This is the USM Strategic Initiative Project for 2021: Talent Management and Wellbeing of USM Staff, in line with the UN SDG3 and SDG4.
>
> The ceremony was officiated by the Vice-Chancellor of USM, Prof. Dato' Dr. Faisal Rafiq Mahamd Adikan in handing over the tablets to several representatives of the recipients.
>
> This project is a collaboration with Kesatuan Kakitangan Am (KKA) USM, with 1260 units to be handed out to the selected recipients.

May this initiative bring benefit to all recipients. Also present at the ceremony were the USM Registrar, Dr. Musa Ali; KKA President, Mr. Ahmad Rizwan Zulkharnain and other key university officials.

*SDG3 – Kesihatan dan Kehidupan yang Sejahtera (Good Health and Well-Being)

SDG4 – Pendidikan Berkualiti (Quality Education)

#usm #wellbeing #b40 #WeLead #SDG3 #SDG4

The post touches on economic and social sustainability themes for the B40 community at the university. It illustrates an example of relatability via the stories it tells to inform the stakeholders of its humanitarian efforts. Positive words such as "donation", "ceremony", "strategic", "initiative", "wellbeing", "collaboration", and "benefit" give readers a feel-good, progressive, charitable vibe. The present tense ("is a collaboration") and past tense ("was officiated") are used, indicating that, although the ceremony has taken place, the collaboration between the parties continues. The post is written in a passive voice that detaches it from its audience. The lack of a first-, second-, or third-person pronoun further distances the message from the stakeholders, making it impersonal. The formal tone resulted in low engagement, with 314 views/likes, five comments, and 12 shares. More personal and inclusive posts are recommended to boost community engagement with USM's Facebook posts. Relatable sustainability stories written in an active tone and using inclusive pronouns and the present tense may encourage higher engagement and physical participation from the audience. The audience should be made to feel that they belong to the community, and that their voice and contribution matter. This may be achieved by using deliberate and more interpersonal grammar and vocabulary to frame a message.

Conclusion

Existing literature has established low ecological awareness and non-green lifestyles among Malaysian citizens, despite various environmental campaigns at state and national levels (Ogiemwonyi et al., 2019). This is indeed worrying, as conspicuous consumption results in depleted resources, continuous pollution, and climate change. It is imperative that Malaysians are made aware of the consequences of their actions for the environment. An effective way to do this is via impactful, continuous sustainability messages on social media from governments, corporations, popular brands, academic institutions, influencers, and NGOs. Steady positive reinforcement of ecological information sharing and campaigns may influence the public to act more sustainably.

As an influential institution, USM can play its role in encouraging desired sustainable behaviours among its community via social media communications. The Facebook page affords ample opportunities to inform, explain, inspire, stimulate, encourage, and motivate contextualised sustainability initiatives in two-way communications and continuous dialogues to produce resilient, aware, and participative stakeholders. This change takes time, as all sustainability messages have deep ideological roots that are influenced by individual experience, geography, history, and culture. Thus, the interpretation of nature is constructed via social interaction that differs from culture to culture (Corbett, 2006, p. 6). As humans and the environment are central to any social change activities, they function within their immediate milieu. When a product's sustainability attributes are aligned with humans' biospheric and altruistic values, they will regard it as relevant and trustworthy. In the long run, a personal norm is stimulated and formed, where communities ascribe sustainability responsibility to themselves and change their behaviour (Tölkes, 2020). Therefore, an emphasis on local systems, norms, and cultures that shift from solely Western-led development are crucial to developing sustainable change (Servaes & Lie, 2015). In this case, USM must act as the local sustainability point for its community, with aims that are aligned with government policy and the 17 SDGs.

Embodying its mission as the proponent to improve the livelihood of the bottom billion, USM's sustainability communication on its Facebook page emphasises its economic and social initiatives more than environmental effort. At its core, USM is determined to transform and develop education, socio-economic status, student mobility, external collaborations, and well-being. Apart from preserving ecosystems, USM develops human and intellectual capital via teaching, learning, and transdisciplinary research across economic development, food production, resource management, sociocultural dimensions, and lifestyle (Foo, 2013). Moving forward, USM may consider balancing the three themes of sustainability (environment, society, economy) in its communications on Facebook. This also relates to USM communities' positive reaction to environment-related posts (e.g., "universiti SAKURA Malaysia"), which indicates their awareness of nature and preservation of its beauty. Indeed, the younger generation of consumers should be targeted to adopt a more ecological lifestyle and become agents of change to solve ecological crises as they are the drivers of the future (Ogiemwonyi et al., 2019).

Importantly, persuasive elements in the Facebook messages should be strengthened to forge a robust connection with the USM communities, so that they respond to, and strive to achieve, a common goal. Compelling storytelling for sustainability or "SusTelling" (Fischer et al., 2020) constructed with a suitable structure and careful choices of lexicogrammar may ensure a more inclusive tone that not only communicates knowledge and issues but also rallies initiatives, participation, and action from the university's communities. The choice of personal pronouns and active constructions may create a more interpersonal, inclusive relationship that coaxes behavioural change and persuades readers that an organisation's image is responsible (Herrick & Pratt, 2013; Nwagbara & Belal, 2019). Stakeholders should be made aware of the university's sustainability endeavours and be informed of the ways they could contribute. Future research may expand the list of sustainability-related words and phrases to include economic and social sectors.

References

Abd Hamid, I., & Wan Yahaya, W. A. (2020). Zero-waste campaign: Assessment on university student's behaviour, awareness, and impact on plastic products. *Malaysian Journal of Social Sciences and Humanities (MJSSH)*, *5*(3), 24–29.

Abd Majid, N., Razman, M. R., Syed Zakaria, S. Z., Ahmed, M. F., & Zulkafli, S. A. (2021). Flood disaster in Malaysia: Approach review, causes and application of geographic information system (GIS) for mapping of flood risk area. *Ecology, Environment and Conservation*, *27*(May), 1–8.

Austin, J. L. (1976). *How to do things with words*. Oxford: Oxford University Press.

Bernier, A. (2020). Sustainability storytelling is not just telling stories about sustainability. *Encyclopedia of the World's Biomes*, *5*, 430–437.

Bhatia, M. (2018). *What is sustainability communication? Is it different than ESG communication? theSustainability.ioTM*. Accessed from https://www.thesustainability.io/what-is-sustainability-communication-esg

Brown, M. (2013). A methodology for mapping meanings in text-based sustainability communication. *Sustainability*, *5*, 2457–2479.

Byrum, K. (2019). "Hey Friend, Buy Green": Social media use to influence eco-purchasing involvement. *Environmental Communication*, *13*(2), 209–221.

Corbett, J. B. (2006). *Communicating nature: How we create and understand environmental messages*. Washington: Island Press.

Cox, M. (2008). Sustainable communication: A study of green advertising and audience reception within the growing arena of corporate social responsibility. Case study: British Petroleum. *Earth & E-Nvironment*, *3*, 32–51.

Crilly, D., Hansen, M., & Zollo, M. (2016). The grammar of decoupling: A cognitive-linguistic perspective on firms' sustainability claims and stakeholders' interpretation. *Academy of Management Journal*, *59*(2), 705–729.

Fernández-Vázquez, J. S. (2021). Analysing the environmental websites of the world's greatest polluters: A multimodal ecolinguistic approach. *Economic Research-Ekonomska Istrazivanja*, *34*(1), 2692–2711.

Fernández-Vázquez, J. S., & Sancho-Rodríguez, Á. (2020). Critical discourse analysis of climate change in IBEX 35 companies. *Technological Forecasting and Social Change*, *157*(May), 120063.

Fischer, D., Selm, H., Sundermann, A., & Storksdieck, M. (2020). Storytelling for sustainability: A definition and its implications for education and communication. In Molthan-Hill, P., Baden, D., Wall, T., Puntha, H., & Luna, H. (Eds.), *Storytelling for Sustainability in Higher Education: An Educator's Handbook* (38–51). London: Routledge.

Font, X., Elgammal, I., & Lamond, I. (2017). Greenhushing: The deliberate under communicating of sustainability practices by tourism businesses. *Journal of Sustainable Tourism*, *25*(7), 1007–1023.

Foo, K. Y. (2013). A vision on the role of environmental higher education contributing to the sustainable development in Malaysia. *Journal of Cleaner Production*, *61*, 6–12.

Genç, R. (2017). The importance of communication in sustainability & sustainable strategies. *Procedia Manufacturing*, *8*, 511–516.

Halliday, M.A.K. (1977). Text as semantic choice in social contexts. In van Dijk, T. A. & Petöfi, J. S. (Eds.), *Grammars and descriptions (studies in text theory and text analysis)*. Berlin: Walter de Gruyter.

Hart, C. 2014. *Discourse, grammar, and ideology: Functional and cognitive perspectives*. London, UK: Bloomsbury.

Herrick, C. N., & Pratt, J. L. (2013). Communication and the narrative basis of sustainability: Observations from the municipal water sector. *Sustainability, 5*(10), 4428–4443.

Huang, L., Clarke, A., Heldsinger, N., & Tian, W. (2019). The communication role of social media in social marketing: A study of the community sustainability knowledge dissemination on LinkedIn and Twitter. *Journal of Marketing Analytics, 7*(2), 64–75.

Ibrahim, J. (2022). Environment "needs better protection." *The Star.* Accessed from https://www.thestar.com.my/news/nation/2022/01/04/environment-needs-better-protection

Ireland, M.E., & Pennebaker, J.W. (2010). Language style matching in writing: Synchrony in essays, correspondence, and poetry. *Journal of Personality and Social Psychology, 99,* 549–571.

Jha, A. K., & Verma, N. K. (2022). Social media sustainability communication: An analysis of firm behaviour and stakeholder responses. *Information Systems Frontiers,* March, 1–20.

Johannessen, K. S., Meldere, I., Torvik, J. M., Aas, T. V., Krasnovs, A., Balteiro, B., Cardoso, R. M. C., & Kraze, S. (2010). The language of advertising: Powerful, innovative and environmental? *New Perspectives on Sustainability, 2,* 6–27.

Johnson, T., & Greenwell, M. (2020, November 12). *Is sustainability advertising just a public relations stunt?* Accessed from https://osf.io/preprints/socarxiv/avy4d/

Kapoor, P. S., Balaji, M. S., & Jiang, Y. (2021). Effectiveness of sustainability communication on social media: Role of message appeal and message source. *International Journal of Contemporary Hospitality Management, 33*(3), 949–972.

Kronrod, A., Grinstein, A., & Wathieu, L. (2012). Go green! Should environmental messages be so assertive? *Journal of Marketing, 76*(1), 96–102.

Lee, C. E., & Chuan, T. L. (2012). Communicating environmental sustainability in Malaysia: A case study. *International Journal of Sustainable Development, 3*(3), 67–76.

Lyon, T. P., & Montgomery, A. W. (2013). Tweetjacked: The impact of social media on corporate greenwash. *Journal of Business Ethics, 118*(4), 747–757.

Martin, J.R., & Rose, D. (2007). *Working with discourse: Meaning beyond the clause* (2nd ed.). London: Continuum.

Mefalopulos, P. (2005). Communication for sustainable development: Applications and challenges. In In Hemer, O. & Tufte, T. (Eds.), *Media and glocal change: Rethinking communication for development* (pp. 247–260). Argentina: Latin American Council of Social Sciences.

Mohd Abdul Kadir, J., Mohmad Hassan, N. N. N., & Abd Aziz, N. N. (2020). Investigating students' attitude and intention to use biodegradable drinking straw in emerging country. *International Journal of Science and Research (IJSR), 9*(1), 418–425.

Mullet, D. R. (2018). A general critical discourse analysis framework for educational research. *Journal of Advanced Academics, 29*(2), 116–142.

Newman, M. L., Pennebaker, J. W., Berry, D. S., & Richards, J. M. (2003). Lying words: Predicting deception from linguistic styles. *Personality and Social Psychology Bulletin, 29*(5), 665–675.

Ndlovu, S. D. (2020). *The extent and nature of green advertising in magazines in South Africa: A content analysis.* [Unpublished doctoral dissertation]. University of Kwazulu-Natal, South Africa.

Nik Hasan, N. N., & Dauda, S. (2020). Roles of Malaysian online newspapers in the construction of public opinion on rare earth risks. *Asian Journal for Public Opinion Research, 8*(4), 432–452.

Nwagbara, U., & Belal, A. (2019). Persuasive language of responsible organisation? A critical discourse analysis of corporate social responsibility (CSR) reports of Nigerian oil companies. *Accounting, Auditing and Accountability Journal, 32*(8), 2395–2420.

Ogiemwonyi, O., Harun, A. B., Othman, B. A., Ismael, D. A., & Ali, R. (2019). Key issues and challenges of green consumer in consuming green product an insight from the emerging country: Malaysia. *International Journal of Psychosocial Rehabilitation, 23*(2), 514–528.

Peirson-Smith, A., & Evans, S. (2017). Fashioning green words and eco language: An examination of the user perception gap for fashion brands promoting sustainable practices. *Fashion Practice, 9*(3), 373–397.

Pranugrahaning, A., Donovan, J. D., Topple, C., & Masli, E. K. (2020). Corporate sustainability assessments in the information communication technology sector in Malaysia. *Sustainability, 12*(21), 1–23.

Rahmat, N.E., Randawar, D.K., Jayabalan, S., Abdul Razak, M.U., & Azmi, I.I. (2021). Revisiting the laws and policies related to educational rights of stateless children in Malaysia. *International Journal of Academic Research in Progressive Education & Development, 10*(3), 1174–1184.

Saadatian, O., Lim, C. H., Mat, S. B., & Sopian, K. (2012). Perspective of sustainable development in Malaysia. *International Journal of Energy and Environment, 6*(2), 260–267.

Saran, S. M., & Shokouhyar, S. (2021). Crossing the chasm between green corporate image and green corporate identity: A text mining, social media-based case study on automakers. *Journal of Strategic Marketing.* DOI: 10.1080/0965254X.2021.1874490.

Sayyed Hussain, S., Mohamad Ali, A., Mohd Kasim, Z., & Jalaluddin, I. (2020). A review on the rhetorical structure and linguistic features of corporate annual reports. *International Journal of Academic Research in Business and Social Sciences, 10*(9), 236–256.

Servaes, J., & Lie, R. (2015). New challenges for communication for sustainable development and social change: A review essay. *Journal of Multicultural Discourses, 10*(1), 124–148.

Stibbe, A. (2009). The language of sustainability: Shouting but not being heard. Keynote address at the Language of Sustainability: Shouting but not being Heard Conference on 27 April 2009, Institution for Environmental Sciences, London.

Tiago, F., Gil, A., Stemberger, S., & Borges-Tiago, T. (2021). Digital sustainability communication in tourism. *Journal of Innovation and Knowledge, 6*(1), 27–34.

Tölkes, C. (2020). The role of sustainability communication in the attitude–behaviour gap of sustainable tourism. *Tourism and Hospitality Research, 20*(1), 117–128.

Valentine, O. (2020). *Social media's influence on green consumerism.* ICDM Pulse. Accessed from https://pulse.icdm.com.my/article/social-medias-influence-on-green-consumerism/

Veland et al. (2018). Narrative matters for sustainability: The transformative role of storytelling in realizing 1.5OC futures. *Current Opinion in Environmental Sustainability, 31*, 41–47.

Vollero, A., Yin, J., & Siano, A. (2022). Convergence or divergence? A comparative analysis of CSR communication by leading firms in Asia, Europe, and North America. *Public Relations Review, 48*(1), 102142.

Weder, F., Lemke, S., & Tungarat, A. (2019). (Re)storying sustainability: The use of story cubes in narrative inquiries to understand individual perceptions of sustainability. *Sustainability, 11*, 5264.

Yeldar, R. (2022). *Words that work.* Accessed from https://ry.com/thinking/words-that-work

3

THE ROLES OF THE SHARING ECONOMY IN PROMOTING SUSTAINABILITY COMMUNICATION AND PRACTICES IN CHINA

Zhang Tong, Huang Miao, and Mohamad Saifudin Mohamad Saleh

Introduction: Sustainability Communication in China

Environmental degradation has led to a significant increase in public concern since the 1960s (Cho et al., 2021). These concerns are widely shared, including by businesses, governments, and non-profit organisations. Defined as "development that meets the needs of the present without compromising the ability of future generations to meet their own needs" (Brundtland Commission, 1987), sustainable development was introduced to balance economic growth with environmental protection. With the necessity to engage in sustainable development, the efforts of organisations to employ sustainability communication have manifested in a variety of ways (Signitzer & Prexl, 2008). Sustainability communication applies to different organisations across multiple domains; it is defined as "an organisation's voluntary, planned, and strategic communication efforts for working towards a balance of economic, social and environment and values in order to achieve the long-term goals of an organisation and its stakeholders" (Ki & Shin, 2015). There are problems with sustainability communication in different countries. For example, a survey of the German market revealed that the public complained about the lack of information on sustainable development (43%) and demanded that sustainable products be made more visible and accessible (42%; Tölkes, 2020).

Under the initiative of the Chinese government, Chinese organisations have gradually adopted sustainability communication as a development goal and put it into practice (Yang et al., 2021). With the development of mobile internet technology, the sharing economy has achieved explosive growth in China (Ping & Xi, 2020). The sharing economy also facilitates the sustainability communication of organisations. This chapter aims to discuss the impact of the sharing economy on sustainability communication in China.

DOI: 10.4324/9781003309642-3

The Sharing Economy in China

The sharing economy is collaborative consumption founded on ecological concepts (Demailly & Novel, 2014). It is a consumption model developed based on the Internet of Things technology (Halim & Hutagalung, 2022). It makes full use of idle resources, meets people's diverse needs, and stimulates economic vitality (Zhang, 2021). The consumption mode of the sharing economy was proposed by Felson and Spaece (1978) and refers to the establishment of a consumption pattern that satisfies individual needs and connects one's own with others' needs under the impetus of ecological civilisation movement. Its key features include an information technology-based marketplace platform created by a third party. That third party can be a business, an organisation, or a government (Felson & Spaeth, 1978).

The theory and practice of the sharing economy originated in the United States and other countries but has developed well in China owing to the implementation of the public ownership economy (Yu & Shen, 2019). The development strategies of the Chinese sharing economy can be divided into the following categories: product sharing (shared bikes, cars, charging banks), space sharing (shared house, shared short-term rental), knowledge and skills sharing (Zhihu and Quora), labour sharing (home delivery platforms), and fund sharing (Liu et al., 2020). Although the forms of the sharing economy are different, they have many commonalities. First, the sharing economy is based on the development of mobile internet technology; second, with the popularisation of the concept of sustainability, people's awareness of environmental protection has improved; third, there are too many idle resources; fourth, the social security credit system gradually improved; and last, the virtual economy has legal and security guarantees (Liang, 2017).

According to a report released by the Chinese government, the sharing economy has been developing rapidly in China since 2017, with the transaction volume of the sharing economy market reaching about 3.69 trillion CNY in 2020 (Xinhua, 2022). The reasons for the rapid development of the sharing economy in China are as follows: (1) the rapid development of information technology is the foundation of the sharing economy, and Liao and Li (2019) point out that the development of productivity, especially the wide application of information technology, has given rise to the sharing economy, a new mode of cooperation, which is also a kind of productivity; (2) the development of the sharing economy is conducive to an increase in the Chinese economic growth rate, the emergence of the sharing economy can increase employment and labour rates, and it is conducive to realising the maximum rate of return on investment (ROI; Sun & Zhang, 2017); (3) the sharing economy promotes sustainability communication in China. Based on an analysis of the connotations, elements, and characteristics of the sharing economy, Yang and Deng (2019) propose that the sharing economy helps to reduce resource consumption.

In addition, there are some problems in the development of the sharing economy in China. First, laws and regulations have not been adapted to the development of the sharing economy. In other words, the Chinese government has not promulgated laws and regulations for the sharing economy. Ma (2016) found that many sharing economy enterprises are suspected of "violations" and face the risk of being punished. Second, there are loopholes in China's credit system, and the sharing economy has changed the situation of traditional economic information asymmetry. Luo and Zhong (2018) believe that the inherent defects in the sharing economy credit mechanism led to uncertainty about consumer protection and transaction security, which is also a problem that urgently needs to be solved. Third, the sharing economy has a crowding-out effect on the traditional economy. Hao's (2019) analysis showed that, in the sharing economy model, users can obtain the right to use products and services for only a small fee and do not need to bear management and maintenance costs, which has a negative impact on traditional business models.

The Sharing Economy and Sustainability Dissemination

At present, 5G, artificial intelligence, big data, cloud computing, and other technologies have caused the transformation of media (Zheng et al., 2016). The internet technology structure, characterised by discreteness and multiple nodes, is at the root of the decentralisation of mobile internet dissemination, which has led to the diversification of information production and dissemination (Sawhney & Suri, 2014). Information dissemination has changed from the one-way, linear dissemination of the past to networked, "explosive" (one-to-many-to-many-to-many) dissemination (Xu & Zhang, 2017). With the development of instant messaging and social media on the mobile internet, the communication structure has changed from "allocation" to "consultation", "registration", and "conversation" (Van Dijk, 2012). As the threshold for use of the mobile internet is reduced, the identity boundary between the disseminator and the recipient is gradually eliminated (Guo, 2021). When the mass media transform into decentralised communication networks, senders become receivers, and producers become consumers (Capriotti, 2017). Therefore, every individual on the mobile internet can produce and disseminate information, while the disseminator status of professional media is weakened, and the ability to set issues is diluted.

The concept of sustainable development needs to rely on effective communication channels to communicate (Janoušková et al., 2019). Sustainability reports and action plans have been published for 30 years, but policymakers and proponents have yet to make the concept popular with a wide range of potential stakeholders (Hák et al., 2018). The essence of sustainable development is a multidisciplinary and interdisciplinary concept. To accomplish this task requires talents in various fields and continuous promotion, not only experts and journalists in relevant fields, but also storytellers (Janoušková et al., 2019). Policymakers would not only plan based on top–down media but would also be influenced by bottom–up media engagement (Servaes et al., 2012). Therefore, it is an effective attempt to instigate as many users on the internet as possible to become communicators.

Driven by mobile internet technology, self-media have developed rapidly, and user-generated content has become the main feature of Web 2.0. Grabs et al. (2018) describe Web 2.0 as the possibility for each user to create content and share with others through various channels. In this context, the key characteristic of user-generated content is that individuals can publish content to others (Yang et al., 2019). User-generated content is content that users have published on a social media platform (Sawhney & Suri, 2014). The term social media includes platforms that contain user-generated content. The development of mobile internet technology has made it easier to publish content on social media platforms (Bala & Verma, 2018).

User-generated content acts as a cognitive trigger and an affective trigger to influence user-responsible environmental behaviour (Sultan et al., 2020). Internet users discussing environmental issues through user-generated content would strengthen their personal emotions and reduce their psychological distress caused by climate change (Elgaaied-Gambier & Mandler, 2021). Internet users browse the content they want to know, create content, form user-generated content, and share selflessly over the internet, thereby encouraging interaction and collaboration among users (Reyes-Menendez et al., 2020), which is the organisational basis of the sharing economy.

The sharing economy integrates idle resources and offline services through online communication and obtains the right to use idle resources with a higher cost-effectiveness (Zhang, 2021). This needs to be completed by connecting online and offline. Neither suppliers nor requesters are commercial organisations; they both transfer and use resources by renting and borrowing to form a commercial transaction (Belk, 2014). Therefore, it is necessary to create sharing platforms to match a large number of requesters and suppliers. Sharing platforms have the ability to collect a large amount of idle resource information. After a certain amount had been reached, it would attract even more requesters and suppliers. They communicate on the platform, share, and form a cycle. The emergence of the platform is the result of the need for development of the sharing economy and the result of recentralisation (Sheng et al., 2018).

As the sharing economy relies on the mobile internet, it shows the characteristics of sustainability communication, such as promoting information sharing, interactivity, and systematic processing (Sheng et al., 2018). Specifically, the owner of resources is the communicator; the content of the sustainability communication is all kinds of information related to sustainability; the specific service content, service carrier, and sustainable information platform are the media; the users of the service are the audiences; and the indicators such as user evaluation of the service and user engagement reflect the sustainability communication effect (Belk, 2014).

Hellobike is a sharing bicycle company founded in September 2016 and headquartered in Shanghai (Yan & Lu, 2019). In August 2021, Hellobike announced that it would build a personal carbon credit system for registered users and creatively use scientific methods in its application to create a carbon account for each user to record carbon emission reductions (Yan & Lu, 2019). The users' cycling mileage is converted into carbon points (Figure 3.1). After a certain number of points have been accumulated, users can win different types of public welfare badges or exchange them for travel benefits. There is also an interactive platform for low-carbon activities, advocating users choose shared bikes to travel. The users are able to share their accumulated carbon points and badges on the platform and invite friends to participate in low-carbon activities to get additional rewards (Figure 3.2).

For the sharing economy, encouraging disseminators to actively publish sustainability information on sharing platforms is the first step in promoting sharing activities (Zhu & Zhou, 2018). From the perspective of the disseminators, they must first ensure the integrity of the sustainability information (Monteiro et al., 2020). Second, they must complete the information exchange, feedback, and confirmation with audiences in the shortest possible time and then provide high-quality, low-cost services to earn favourable comments from audiences so that their idle resources continue to create value (Turna, 2022).

FIGURE 3.1 Hellobike carbon account page.

FIGURE 3.2 Hellobike interactive platform page.

Two-way Sustainability Communication in the Sharing Economy

In traditional markets, information asymmetry involves the study of transaction decisions in which one party has more or better information than the other, creating an imbalance of power in transactions where the party with more information gets more benefit (Lu et al., 2019). The mobile internet is a tool for sustainability communication and processing, but its emergence has not completely solved the problem of information asymmetry. Information asymmetry is an objective fact. In the dissemination of information on the internet, information asymmetry often occurs owing to the concealment of the networks of senders and receivers; the amount of knowledge possessed by them is not equal, and the channels for obtaining information are different. Asymmetric information is mainly manifested in the unbalanced distribution of information among the information subjects, resulting in one party having more information than the other party.

Solving climate and sustainability communication issues in society requires more dialogues, debates, and investigations (Pompper, 2016). Divergence on climate change requires more attention being paid to its crux than to direct action (Hennes et al., 2016). Social media with two-way communication capabilities should be given more attention as a medium to persuade citizens and pressure opinion leaders to adopt positive and sustainable communication (Pompper, 2016). Mass media and other forms of communication act not only as a one-way form of self-promotion, but as a facilitator for two-way communication between stakeholders to understand and solve problems (Pompper, 2016).

In the sustainability communication of mobile internet dissemination, there are two main information asymmetries. One is the asymmetry caused by the lack of connection of information obtained by both parties, which means that, when requesters are looking for sustainability information or knowledge, they do not know how to send information or how to spread the information. As a result, the supplier of information does not know it, and, thus, the information cannot be provided in time. Another asymmetry is that people who have a lot of information are unwilling to provide it (Buda et al., 2019). As long as there is information asymmetry between suppliers and requesters, defective products will appear (Akerlof, 1978). Sharing platforms are able to effectively avoid the appearance of defective products through efficient information communication and direct two-way communication (Akerlof, 1978; Lu et al., 2019).

The two-way communication enabled by the mobile internet has realised equality of information and efficient sustainability communication (Lu et al., 2019). Compared with traditional methods, users are able to communicate in more aspects and get feedback. The interactivity of mobile communication makes the identities of the two parties constantly change, breaking boundaries, weakening the strong position of the communicator and the possibility of the audience being manipulated, and disintegrating the unbalanced pattern. In order to reduce information asymmetry and allow users to communicate directly and effectively, sharing platforms change the way users "connect" (Zhu et al., 2019).

With a traditional hotel reservation method or on a hotel reservation website, the user usually browses and selects the desired room and submits a reservation request (Midgett et al., 2018). After receiving the order, the third-party intermediary platform processes it, sends the order to the reserved hotel, and then waits for the hotel to confirm and send a message to the user to complete the reservation (Sung et al., 2018). With the support of internet technology, it only takes a few minutes for these steps to be transmitted through the network, but, during the whole process, there is no communication between the user and the hotel (Midgett et al., 2018). There might be potential problems during this period, such as users with special requirements (such as adding supplies, late arrivals, or room preferences). And, because the hotel has more information, but the user does not, such information asymmetry might cause defective products (Akerlof, 1978).

As one of the typical representatives of a sharing economy platform, Airbnb's booking system breaks this traditional model (Narasimhan et al., 2018). When users browse and find the room they need, they can click to communicate directly with the host and ask for information about the room, such as the distance from the train station and geographic location information, and can also make specific requests (Sung et al., 2018). The host knows everything about the trip and destination from the users' description and can finally decide whether or not to rent it out. Two-way direct communication enables effective feedback and better exchange of information between both sides (Lu et al., 2019). During the whole process, Airbnb plays the role of a platform, handing over the right of communication to users and hosts (Narasimhan et al., 2018). This transforms the traditional third-party "connection" into a connection method for direct communication between users and hosts (Zhu et al., 2019). The communication between users and hosts is different from general website bookings: it resembles communication between friends and is more in line with the trend of personalised customisation (Midgett et al., 2018). The host and the user are not known to each other, but, after communication, the rental experience is more like visiting a friend's house. Airbnb also expands the depth of connection (Zhu et al., 2019).

Airbnb focuses on using the social features of the sharing platform to enhance users' experience (Midgett et al., 2018). Airbnb understands demands from the user's point of view. After collecting users' feedback, it found that users pay attention to the authenticity of images, and it hires photographers to provide professional photographs to ensure that the visual effects are improved under real conditions (Narasimhan et al., 2018). These images can be one of the important reasons that attract users to choose Airbnb homes (Sung et al., 2018). The sharing of users' rental experiences on social media has increased the number of users joining sharing platforms and effectively promoting sustainability communication with the public. In addition, Airbnb advocates the concept of green tourism (Zhu & Zhou, 2018). Airbnb advises hosts to use eco-friendly cleaning products, install solar power equipment, and encourage tenants to use public transportation, while meeting basic lodging functions (Wei, 2018). Sixty-six per cent of renters said the positive impact of shared accommodation on the environment is an important factor in their choice of Airbnb (Wei, 2018). By connecting with users on the sharing platform, Airbnb is able to accurately deliver sustainability information and improve sustainability communication.

The Advantages of the Sharing Economy in China: Promotion of Sustainability Communication

As the largest market in the world, China is currently facing an economic transformation. Sustainability is an urgent issue for China owing to the heavy pressure industrialisation puts on society and resources (Shi & Zhang, 2006). Given the gradually growing shortage of various resources, the sharing economy, as an intensive and efficient low-carbon economy, is the future development direction of the Chinese economy (Cai et al., 2013). China has advantages in the rapid construction of mobile internet infrastructure and the number of internet users (Ren et al., 2021). These advantages of the sharing economy would promote sustainability communication in China.

Mobile communication infrastructure is one of the important public infrastructures for China's economic and social development (Chen & Kang, 2018).

The construction of 5G networks is advancing rapidly. China has installed 1.425 million 5G base stations, which form the world's biggest 5G network (Deng, 2022). The construction of information infrastructure is a guarantee for rapid dissemination of information (Zeng & Lo, 2020). People's lifestyles and communication methods have also changed owing to the popularisation of the mobile internet, and the efficiency of sustainability communication has also improved (Zeng & Lo, 2020).

With the wide coverage of the mobile internet, the amount of internet users in China has increased year by year. The number of internet users has exceeded 1 billion, and the proportion of internet users using mobile phones has reached 99.7%, which indicates the huge potential of mobile internet communication (*Global Times*, 2022). At present, China's sharing economy is mainly concentrated in the fields of finance, life services, transportation, production capacity, knowledge and skills, and short-term housing rental, but it will expand into other fields in the future (Zhu & Zhou, 2018). The huge number of mobile users produces more idle resources, and this capacity could be integrated into sharing platforms and disseminated to more stakeholders (Zhu & Zhou, 2018).

One of the best examples of local transportation in China is represented by the ride-hailing start-up company DiDi, a company headquartered in Beijing that provides app-based transportation services, including taxi-hailing, private car-hailing, social ride-sharing, and bike-sharing (Giovannini, 2020). DiDi highlights the most environmentally friendly option when choosing a way to travel and explains in text that the option would reduce the amount of carbon emissions; users who choose options with lower carbon emissions would get more discounts. Users are more likely to choose environmentally friendly options when attracted by low prices (Giovannini, 2020). Meanwhile, DiDi taxis have eco-friendly stickers on their body, and the benefits to sustainable development of sharing trips are posted on the backs of seats.

Conclusion

As an efficient, low-carbon economy, the sharing economy represents the direction of future development, but the essence of a sharing economy is communication activity. The decentralisation of the sharing economy leads to an increase in user-generated content and users' preference for sharing content on social media, expanding the spread of sustainability communication. In addition, two-way mobile communication realises equal dissemination of information, whereby users can communicate with each other directly and get feedback instantaneously. Sustainability communication based on mobile dissemination removes unbalanced dissemination and gives users equal rights to obtain information. The sharing economy stimulates the audience's initiatives as users who are eager to obtain more sustainable information. When this demand is met, better two-way communications and sustainability communications can be realised and achieved.

References

Akerlof, G. A. (1978). The market for "lemons": Quality uncertainty and the market mechanism. In *Uncertainty in economics* (pp. 235–251). Academic Press.

Bala, M. & Verma, D. (2018). A critical review of digital marketing. *International Journal of Management, IT & Engineering, 8*(10), 321–339.

Belk, R. (2014). You are what you can access: Sharing and collaborative consumption online. *Journal of business research, 67*(8), 1595–1600.

Brundtland Commission. (1987). *Our common future. The Brundtland Report.* Oxford, UK: World Council on Sustainable Development.

Buda, G., Pethes, B., & Lehota, J. (2019). Dominant consumer attitudes in the sharing economy - A representative study in Hungary. *Resources, 9*(1), 1.

Cai, S., Chen, X., & Bose, I. (2013). Exploring the role of IT for environmental sustainability in China: An empirical analysis. *International Journal of Production Economics, 146*(2), 491–500.

Capriotti, P. (2017). The world wide web and the social media as tools of CSR communication. *Handbook of integrated CSR communication* (pp. 193–210). Cham: Springer.

Chen, S. Z., & Kang, S. L. (2018). A tutorial on 5G and the progress in China. *Frontiers of Information Technology & Electronic Engineering, 19*(3), 309–321.

Cho, M., Park, S. Y., & Kim, S. (2021). When an organization violates public expectations: A comparative analysis of sustainability communication for corporate and nonprofit organizations. *Public Relations Review, 47*(1), 101928.

Demailly, D., & Novel, A. S. (2014). The sharing economy: make it sustainable. *Studies, 3*(14), 14–30.

Deng, I. (2022, March 9). China targets 2 million installed 5G base stations this year, expanding world's biggest next-generation mobile network, as 6G preparations push ahead. *South China Morning Post.* Accessed from https://www.scmp.com/tech/policy/article/3169832/china-targets-2-million-installed-5g-base-stations-year-expanding

Elgaaied-Gambier, L., & Mandler, T. (2021). Me trying to talk about sustainability: Exploring the psychological and social implications of environmental threats through user-generated content. *Ecological Economics, 187*, 107089.

Felson, M., & Spaeth, J. L. (1978). Community structure and collaborative consumption: A routine activity approach. *American Behavioral Scientist, 21*(4), 614–624.

Giovannini, M. (2020, January 31). DiDi's sustainable transportation investments promote better cities. *CGTN.* Accessed from https://news.cgtn.com/news/2020-01-31/Didi-s-sustainable-transportation-investments-promote-better-cities-NHLqFLvVKw/index.html

Guo, H. (2021). The development and application of new media technology in news communication industry. *The International Journal of Electrical Engineering & Education.* February.

Global Times. (2022, February 25). China has 1.032 billion internet users, 73.0% penetration rate. *Global Times.* Accessed from https://www.globaltimes.cn/page/202202/1253226.shtml

Grabs, A., Bannour, K. P., & Vogl, E. (2018). *Follow me.* Bonn: Rheinwerk Verlag GmbH.

Hák, T., Janouškova, S., Moldan, B., & Dahl, A. L. (2018). Closing the sustainability gap: 30 years after "Our Common Future", society lacks meaningful stories and relevant indicators to make the right decisions and build public support. *Ecological Indicators, 87*, 193–195.

Halim, D. K., & Hutagalung, S. (2022). Towards data sharing economy on Internet of Things: a semantic for telemetry data. *Journal of Big Data, 9*(1), 1–24.
Hao, F. F. (2019). Examination and reflection on sharing economy in China. *Business Economics Research, 06*,190–192.
Hennes, E. P., Ruisch, B. C., Feygina, I., Monteiro, C. A., & Jost, J. T. (2016). Motivated recall in the service of the economic system: The case of anthropogenic climate change. *Journal of Experimental Psychology: General, 145*(6), 755.
Janoušková, S., Hák, T., Nečas, V., & Moldan, B. (2019). Sustainable development - a poorly communicated concept by mass media. Another challenge for SDGs? *Sustainability, 11*(11), 3181.
Ki, E. J., & Shin, S. (2015). Organization sustainability communication (OSC): Similarities and differences of OSC messages in the United States and South Korea. *Computers in Human Behavior, 48*, 36–43.
Liang, W. F. (2017). Research on the strategic transformation of Chinese green employment under the development of sharing economy. *Reform and Strategy 05*,141–143.
Liao, P. P., & Li, J. J. (2019). Research on the sharing economy from the perspective of Marx's cooperation thought. *Southeast Academic, 02*, 47–56.
Liu, C., Chan, R. K., Wang, M., & Yang, Z. (2020). Mapping the sharing economy in China. *Sustainability, 12*(16), 6333.
Lu, D., Lai, I. K. W., & Liu, Y. (2019). The consumer acceptance of smart product-service systems in sharing economy: The effects of perceived interactivity and particularity. *Sustainability, 11*(3), 928.
Luo, Y., & Zhong, G. Y. (2018). Research on government regulation innovation for sharing economy. *Hunan Social Sciences, 02*, 121–126.
Ma, Q. (2016). The development status, bottleneck and countermeasures of sharing economy in China. *Discussion on Modern Economy, 10*, 20–24.
Monteiro, S., Ribeiro, V., & Lemos, K. (2020). Emerging trends on sustainability/integrated reporting: public and private perspectives. *CSR and Sustainability in the Public Sector*, 137–149. Springer: Singapore.
Midgett, C., Bendickson, J. S., Muldoon, J., & Solomon, S. J. (2018). The sharing economy and sustainability: A case for Airbnb. *Small Business Institute Journal, 13*(2), 51–71.
Narasimhan, C., Papatla, P., Jiang, B., Kopalle, P. K., Messinger, P. R., Moorthy, S., Proserpio, D., Subramanian, U., Wu, C., & Zhu, T. (2018). Sharing economy: Review of current research and future directions. *Customer Needs and Solutions, 5*(1), 93–106.
Ping, Z., & Xi, S. (2020). An effective path to made in China 2025 - Sharing economy mode. *German and Chinese Contributions to Digitalization*, 89–97. Springer Gabler: Wiesbaden
Pompper, D. (2016). Beyond the business case: Building upon traditional approaches and opening new spaces for multiple perspectives on climate and sustainability communication. *Mass Communication and Society, 19*(5), 543–547.
Ren, S., Hao, Y., Xu, L., Wu, H., & Ba, N. (2021). Digitalization and energy: How does internet development affect China's energy consumption? *Energy Economics, 98*, 105220.
Reyes-Menendez, A., Saura, J. R., & Filipe, F. (2020). Marketing challenges in the #MeToo era: Gaining business insights using an exploratory sentiment analysis. *Heliyon, 6*(3), e03626.
Sawhney, H., & Suri, V. R. (2014). From hierarchy to open configurations: decentralization and user-generated content. *Media, Culture & Society, 36*(2), 234–245.

Servaes, J., Polk, E., Shi, S., Reilly, D., & Yakupitijage, T. (2012). Towards a framework of sustainability indicators for 'communication for development and social change' projects. *International Communication Gazette, 74*(2), 99–123.

Sheng, Z., Wang, Y., & Sheng, Y. (2018). The Impact of shared economy on the business model: From decentralization to recentralization. In International Conference on Management Science and Engineering Management (pp. 97–108). Springer, Cham.

Shi, H., & Zhang, L. (2006). China's environmental governance of rapid industrialisation. *Environmental Politics, 15*(02), 271–292.

Signitzer, B., & Prexl, A. (2008). Corporate sustainability communications: Aspects of theory and professionalization. *Journal of Public Relations Research, 20*(1), 1–19.

Sultan, M. T., Sharmin, F., Badulescu, A., Stiubea, E., & Xue, K. (2020). Travelers' responsible environmental behavior towards sustainable coastal tourism: An empirical investigation on social media user-generated content. *Sustainability, 13*(1), 56.

Sun, Y. H., & Zhang, G. (2017). Governance in sharing economy. *China Development, 01*, 80–87.

Sung, E., Kim, H., & Lee, D. (2018). Why do people consume and provide sharing economy accommodation? - A sustainability perspective. *Sustainability, 10*(6), 2072.

Tölkes, C. (2020). The role of sustainability communication in the attitude - behaviour gap of sustainable tourism. *Tourism and Hospitality Research, 20*(1), 117–128.

Turna, G. B. (2022). Impact of social media on tourism, hospitality and events. *Handbook on Tourism and Social Media* (pp. 475–488).

Van Dijk, J. A. (2012). *The network society (3rd edition)*. London, UK: Sage Publications Ltd.

Wei, X. N. (2018). *Airbnb's "healthy tourism" initiative: based on improving people's livelihood and focusing on sustainable development*. Visit Beijing. https://www.visitbeijing.com.cn/article/47QsmKTYSP6

Xinhua. (2022). *China's sharing economy market turnover grows 9.2% in 2021*. The state council, the people's republic of China. http://english.www.gov.cn/archive/statistics/202202/27/content_WS621b272fc6d09c94e48a589b.html

Xu, Y., & Zhang, W. (2017). New Media's Impact on China from a Culture Perspective. *In New Media and Chinese Society, 5*, 95–119.

Yan, Z., & Lu, J. (2019). A Comparison of Service between Hellobike and traditional public bicycle in Jingjiang. *CICTP, 2019*, 1410–1419.

Yang, M., Ren, Y., & Adomavicius, G. (2019). Understanding user-generated content and customer engagement on Facebook business pages. *Information Systems Research, 30*(3), 839–855.

Yang, Y., Xu, H., Wang, J., Liu, T., & Wang, H. (2021). Integrating climate change factor into strategic environmental assessment in China. *Environmental Impact Assessment Review, 89*, 106585.

Yang, Z. K., & Deng, C. H. (2019). The impact of sharing economy on the construction of ecological civilization in the new era. *Social Science Front, 02*, 251–255.

Yu, T., & Shen, W. (2019). Funds sharing regulation in the context of the sharing economy: Understanding the logic of China's P2P lending regulation. *Computer Law & Security Review, 35*(1), 42–58.

Zeng, F., Lee, S. H. N., & Lo, C. K. Y. (2020). The role of information systems in the sustainable development of enterprises: A systematic literature network analysis. *Sustainability, 12*(8), 3337.

Zhang, Y. (2021). The goal and essence of sharing economy. In *Sharing economics* (85–102). Singapore: Springer.

Zhang, Y. (2021). Sharing economy business model. In *Sharing economics* (pp. 257–281). Singapore: Springer.

Zheng, K., Yang, Z., Zhang, K., Chatzimisios, P., Yang, K., & Xiang, W. (2016). Big data-driven optimization for mobile networks toward 5G. *IEEE Network, 30*(1), 44–51.

Zhu, G., Li, H., & Zhou, L. (2018). Enhancing the development of sharing economy to mitigate the carbon emission: a case study of online ride-hailing development in China. *Natural Hazards, 91*(2), 611–633.

Zhu, Y., Cheng, M., Wang, J., Ma, L., & Jiang, R. (2019). The construction of home feeling by Airbnb guests in the sharing economy: A semantics perspective. *Annals of Tourism Research, 75*, 308–321.

4

STRATEGIC SUSTAINABILITY COMMUNICATION MODEL OF THE COMPANIES LISTED ON THE STOCK EXCHANGE OF THAILAND

Rittipol Kuntasuwun

Introduction: Sustainability from Thailand's Perspective

The sustainable development movement in Thailand has been documented in a book named *Thailand's Sustainable Development Sourcebook: Issues and Information, Ideas, and Inspiration* by Grossman, Algie, Treerutkuarkul, and Wegner (2015). There are many factors that contributed to the "Bubble Economy" situation, such as population increase, scarcity of natural resources, demand for energy consumption, climate change, income inequality, and the challenges of society, culture, and the environment. Like any other country, its focus on sustainability is rooted in its specific background and features. In Thailand's context, there was a signal from the 1997 economic crisis known as the "Bubble Economy". Over the preceding ten years, Thailand was ranked among the Asian tiger economies, enjoying economic growth, low inflation, foreign investment in infrastructure projects, automobile and electronics industries, textiles and other light industries, as well as great development of real estate and service industries. The growing hospitality and tourism businesses could be identified by the ongoing expansion of department stores, golf clubs, international schools, and hospitals.

The economic crisis in 1997 resulted in the devaluation of the Thai baht currency. In addition, the decrease in import and export investments, high international debts, and trading of stocks without fundamental business analysis resulted in an economic breakdown. The Bank of Thailand spent up to 90% of the kingdom's foreign reserves to defend the Thai currency. In the end, the Thai government had no alternative but to float the currency from 25 baht to more than 57 baht per US dollar. During the crisis, His Majesty the King Rama IX remarked to the Thai people in his annual birthday address in December 1997:

> I have often said to the audience that to be a tiger is not important. The important thing for us is to have a self-reliant economy. A self-reliant

DOI: 10.4324/9781003309642-4

> economy means having enough to survive. That does not mean that you all must produce your own foods or knit your own clothes. I think that is so overwhelming. The area of your residence should have proper sufficiency, if you can produce something more than you demand, you can sell it with low cost of logistics. Some economists might think that it is outdated because they think that the economy must be commercial, not sufficiency. But fortunately, Thailand has the advantage that we can produce to be sufficient.
>
> *(Grossman et al., 2015,* p. 32–33*)*

The king's idea has been developed as the "Sufficiency Economy Philosophy", which would not only be for the cultivation and guidance of Thailand's economic recovery, but also offer immunity to economic crises in the future. Moreover, the king's remark about sustainability overlaps with the concept of Buddhist teachings, to which 95% of Thais adhere. Buddhist teachings focus on a moderate mindset, immaterialism, interdependence, and the importance of cause and effect of actions. The king's idea and Buddhist inspiration have been the encouragement and direction for suitable sustainable development. The philosophy of sufficiency economy could be adapted and applied effectively in the various areas of cultural contexts and social values (Grossman et al., 2015).

Governments and Sustainability

In 2015, the Thai cabinet saw that Thailand had no vision for the country or long-term goals and strategy. There was also discontinuous government administration because of the frequent cabinet changes, which affected the country's adaptation to global paradigm shifts. The reformation of the government administration system should be targeted as a long-term goal; thus Thailand proposed the "20-year National Strategy Framework (2017–2036)" to tackle relevant problematic issues affecting the country from around the world (Office of the National Economic and Social Development Board, 2016).

The 20-year National Strategy stated its vision of "Thailand as a developed nation with Stability, Prosperity, and Sustainability in accordance with the principles of Sufficiency Economy Philosophy (SEP)" and the national motto as "Stability, Prosperity, and Sustainability". The National Strategy is to be transformed into action through the 5-year National Economic and Social Development Plans (NESDB), inaugurating the Twelfth Plan from October 2017 to September 2021.

In order to accomplish the vision of Thailand as a developed nation with stability, prosperity, and sustainability, based on the principles of Sufficiency Economy Philosophy (SEP) or "Stability, Prosperity, and Sustainability", Thailand has to initiate long-term national strategies that will enhance continuous operation and implementation. Public integration and cooperation will accumulate higher income, allowing it to become a developed country. This requires pioneering, competitive competency, possession of stability and equality, and improved quality of life.

This chapter focuses on the "Sustainability" dimension of the National Strategy based on Thailand's roots of sustainable development in the SEP. This SEP theme has integrated many aspects with distinctive characteristics by understanding causes and effects, recognising moderation, and integrating technology, innovation, creativity, and culture. All of these are done to balance and sustain the economy, society, and environment dimension.

The Thai government has supported sustainable development implementation and related projects by establishing the National Committee for Sustainable Development or CSD, chaired by the prime minister. The CSD's main responsibility for the country's sustainable development includes initiating policies and strategies, supporting other related sectors in sustainable development, supervising other government sectors, encouraging the improvement of economics, society, and laws, and following and evaluating the outputs. This committee comprises three subcommittees, which were formed of integral and cooperative representatives from all sectors, namely government, the private sector, nonprofit organisations, and professionals. The three subcommittees focus on mobilising the Sustainable Development Goals (SDGs), raising awareness of sustainable development and the application of the Sufficiency Economy Philosophy, and compiling data and statistics to support the implementation and monitoring of the 2030 Agenda.

Private Sectors and Sustainability

Many international companies have significantly increased interest in "sustainability" issues by focusing on important environmentally friendly manufacturing processes, including operating businesses legally and with responsibility to society and community concerns (Friedman, 2009; Nidumolu, Prahalad, & Rangaswami, 2009; van Marrewijk & Werre, 2003). Many companies integrate sustainable development in their operations and supply chain, along with creating shared value concepts. These would increase awareness and co-creation among stakeholders. Moreover, corporate businesses have realised that sustainable development is a significant opportunity to maximise resource usage, decrease unnecessary expenses, and pursue the concept of creating a corporate image as an environmentally friendly company.

A good corporate reputation also has been derived from corporate social responsibility implementation. It focuses on operating businesses sustainably, which is a significant application in terms of the dramatic paradigm shift in the business world. According to McKinsey's survey focusing on executives, it is found that operating businesses with sustainability concerns is another important factor to create a positive corporate reputation (Bonini, 2012). The Stock Exchange of Thailand (SET) plays an important role as another sector that can drive and execute the mission to encourage Thai companies to pursuing sustainable business growth. The SET has three significant areas of sustainable development, including encouraging stakeholders of the capital market to determinedly focus on sustainability in business processes and investment, fostering the development of the ecosystem to promote sustainability of the capital market, and elevating the sustainable development of the Thai capital market to be globally accepted.

The SET renamed the Corporate Social Responsibility Institute the Social Responsibility Centre, or SR Centre, in 2007. This centre has its own objectives, focusing on formulating sustainable development guidance for listed companies by developing practical knowledge, encouraging sustainability compliance, pursuing implementation, distributing sustainability information, and setting up evaluation procedures for sustainability (The Stock Exchange of Thailand, 2015). In 2015, SET became a member of the UN Sustainable Stock Exchange (UN SSE Initiative), an initiative for trading sustainable stocks. Since then, the SET has continuously promoted and developed sustainability implementation by listed companies by improving tools for knowledge learning and executing sustainability in all supply chain areas. The SET also encourages listed companies to drive transparency through sustainability reports, compliments and offers motivating rewards to listed companies that comply with sustainability procedures. The SET also selects highly competent companies to assist in the evaluation and assessment process for granting SET awards (SET, 2015).

Previously, in 2006, the SET awarded sustainability certification to listed companies with outstanding sustainability performance and recognition from other companies. This award also provided encouragement to executives and professionals in listed companies with strong aspirations to develop their businesses for sustainability (SET, 2015).

In 2015, SET encouraged quality listed companies to vividly reflect sustainable development business operations with the combination of the Top Corporate Governance Report Awards and Best Corporate Social Responsibility Awards. The combined awards are named the Sustainability Awards and have the objectives of certifying outstanding listed companies and encouraging executives and professionals in the sustainability field who have a strong commitment to operating sustainable businesses (SET, 2015).

Before conducting a sustainability evaluation, SET would initiate Thailand Sustainability Investment (THSI) from listed companies that passed the criteria and evaluation measurement in terms of environmental, social, and governance (ESG) dimensions. The announcement of the THSI list would be the showcase and would factor in the investment decision-making process of investors. The other benefits would be to reduce listed companies' burden of preparing data and documents because the information provided in the sustainability questionnaire for the Sustainability Awards would form a significant database listing THSI (SET, 2015).

The significant consideration procedure for the SET Sustainability Awards involves the information provided in the questionnaire including policy, vision, and participation in encouraging corporate sustainability. The consideration focuses on the information from core business operations with internal and external stakeholders' relevance in terms of economic, social, and environmental aspects.

The SET Sustainability Awards are another significant factor stimulating, encouraging, and framing the guidance for listed companies to perform their sustainable business under environmental, social, and corporate governance circumstances. If the companies implemented their sustainable business growth initiatives, it would at least bring stability and prosperity to the stakeholders in each segment of the company. As a result, this would further enhance the country's prosperity.

Listed Companies' Executives and Stakeholders

This chapter explores the strategic sustainability communication model for businesses in order to acquire an overview of, and tendency of direction in, implementation of strategic sustainability communication by companies listed on the Stock Exchange of Thailand. The data have been collected by interviewing management-level executives from ten companies and eight representatives from seven groups of stakeholders. The ten management-level executives of listed companies were selected from five companies that have received SET Sustainability Awards and another five that have not received awards. The author collected data from analysis of documents related to sustainability communication and in-depth interviews for analysis by comparing strategic sustainability dimensions, overviews of listed companies' sustainability implementation, implementation of sustainability guidelines from the SET, sustainability communication strategies, and sustainability communication tactics to reach sustainability goals. These data were integrated into the sustainability overview of companies listed on the SET and implemented as a strategic sustainability communication model.

The names of the representative companies are the PTT Public Company Limited, Minor International Public Company Limited, the Siam Cement Public Company Limited, Thai Oil Public Company Limited, PTT Global Chemical Public Company Limited, Airports of Thailand Public Company Limited, PTT Exploration and Production Public Company Limited, Charoen Pokphand Foods Public Company Limited, and the Siam Commercial Bank Public Company Limited.

Data were also collected from in-depth interviews of eight interviewees who were representatives of the seven groups of stakeholders. These comprised civil society and non-governmental organisations (NGO, government, executive and employee, customer and consumer, shareholder, media, as well as academics and professionals, namely Thai Environmental and Community Development Association, SET, PTT Global Chemical Public Company Limited, Luffala Social Enterprise, Krungsri Asset Management Company Limited, Bangkok

Broadcasting & TV Company Limited, Chiang Mai University, and Thailand Public Relations Association).

The themes derived from the analysis are sustainability communication process, tendency and direction of strategic sustainability communication, and the model of strategic sustainability communication among companies listed on the SET.

The Strategic Sustainability Communication Process among Companies Listed on the SET

The implementation of strategic sustainability communication by companies listed on the Stock Exchange of Thailand focuses on stakeholders as the significant target audience. The stakeholders were grouped by the type of business and industry. Later, the expectations and needs of each stakeholder were analysed to develop content and to select communication tactics as well as channels to match them. The essential objective was to enhance the understanding between stakeholder and company. In terms of the tendency and direction of sustainability communication, all listed companies shared similar opinions regarding stakeholders' orientation. For example, information acquisition, content generation, and communication tactics selection must be appropriate and matched to each type of stakeholder. In terms of communication channel selection, they also mentioned that digital media are crucial and play an important role.

From the above discussion, the strategic sustainability communication process of companies listed on the Stock Exchange of Thailand from the representatives' point of view is presented in Figure 4.1.

Roles of Stakeholders in Relation to Sustainability Communication

Stakeholders have different roles in relation to sustainability communication depending on the person or sectors that they deal with. The stakeholders must

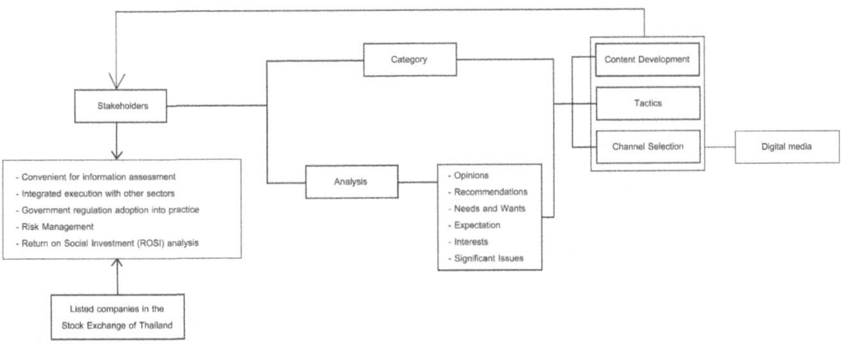

FIGURE 4.1 Thai representatives' view of the strategic sustainability communication process among companies listed on the SET.

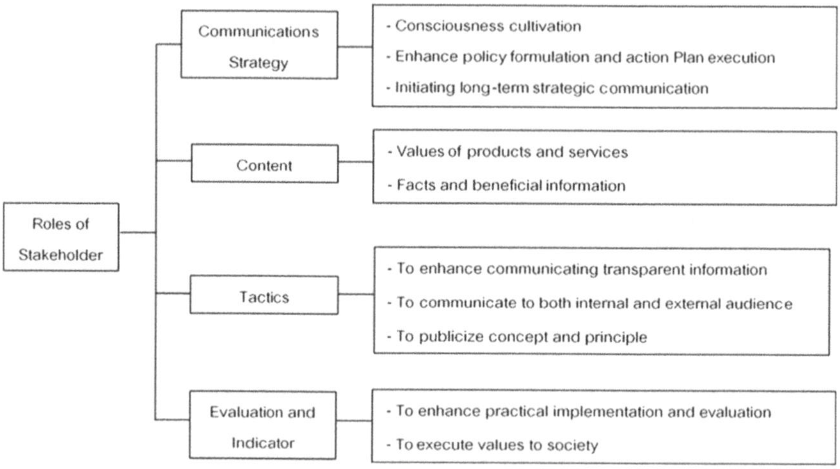

FIGURE 4.2 Roles of stakeholders in sustainability communication.

begin their roles in sustainability policy formulation and action plan execution, as well as investigation of outcomes that derived from communication. The roles can be identified in Figure 4.2.

The roles of stakeholders illustrated in Figure 4.2 are explained in more detail in Table 4.1.

Stakeholders' Expectations of Listed Companies in Relation to Sustainability Communication

Stakeholders expected the companies listed on the Stock Exchange of Thailand to focus on them as the target audience in their sustainability communication effort. In the process, content and information must be appropriate to each group of stakeholders. The stakeholders' expectations can be seen in Figure 4.3.

Stakeholders have voiced their expectations that the companies should offer transparent information with spontaneous responses. Moreover, stakeholders' expectations and needs must be analysed to understand the stakeholders. Stakeholders believe that they are another factor that plays an important role in formulating an initial strategy and connecting sustainability with the relevant business strategy.

Stakeholders suggest that the company should communicate sustainability to progress with business operations. It is necessary for the companies operating businesses to complete impromptu adaptation and improvement, explore new business opportunities, operate with transparency, publish accurate information with regard to regulation concerns, implement CSR activity, as well as respond to customers with high-quality safe products. Should these be reflected upon and

TABLE 4.1 The roles of stakeholders in relation to sustainability communication of companies listed on the SET

Roles in Relation to Sustainability Communication	*Detail of Roles in Relation to Sustainability Communication*
Consciousness cultivation	To communicate sustainability to target audiences by cultivating consciousness in them, so that they can transfer information to sectors with which they interact
Policy and action plan enhancement	To enhance policy formulation and initiate sustainability action plans
To enhance action plans and follow up	To promote sustainability action plans in concrete practice and follow up
To promote transparent information exposure	To communicate transparent and investigable information
To communicate to internal and external audiences	To communicate sustainability to internal and external stakeholders
To communicate values through products and services	To communicate values through products and services by conveying that they will receive products and services with quality, high-standard processing, as well as to create financial sustainability and careers for communities
Long-term communication strategy	To initiate long-term strategic communication to create long-term perception and understanding by developing contents and selecting communication channels appropriate to each group of stakeholders
To communicate facts and beneficial information	To communicate essential facts, especially significant information that directly relates to target
To implement valuable action plans for society	To implement concrete and valuable action plans for society by continuous improvement and communicating to external sectors
To publicise concept and principle	To publicise sustainability concept and principle within related works such as consulting, lectures, research, articles, and studies

implemented, there is a great opportunity for a business to experience sustainable growth and evoke trust among investors.

Business operations and internal management's policy and vision must be communicated to employees and publicised to external sectors with activities or projects. The activities or projects must attract participation from all sectors that are directly involved in carrying out sustainability efforts. A company must create business alliances that could be drawn from participation in activities.

For the communication content, the company must consider sustainability the fundamental concern. The communication tactics need to formulate objectives, contents, communication strategies, tactics selections, outcomes, follow-ups, and improvements. Companies and the media should communicate benefits

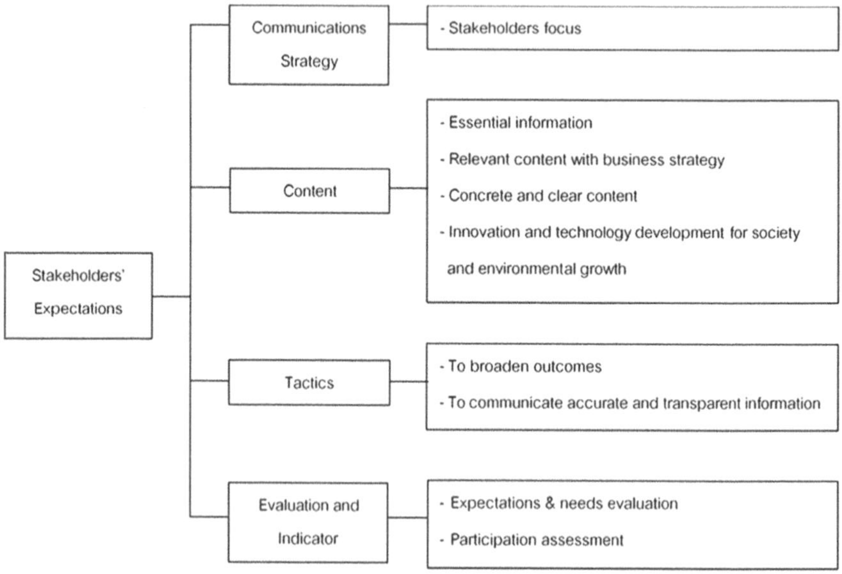

FIGURE 4.3 Stakeholders' expectations for sustainability communication.

and useful data to help audiences develop their opinions based on valid sources instead of questioning the information.

Companies must also pay attention to relevant groups of stakeholders. For example, a company must create trust and show reliability in a group of investors and shareholders to enhance their readiness to invest. Employees are indeed significant communication agents for a company. If the employees understand sustainability, they will have the confidence to communicate it to external sectors. Groups of business partners must perceive the company's transparent content. Finally, as for the community and society, companies must implement sustainability at national, regional, provincial, and local levels. This is to enhance credibility and cultivate transparent communication habits among company stakeholders to create stability and sustainability for businesses.

The Direction of Strategic Sustainability Communication among Companies Listed on the SET

The information analysis based on the in-depth interviews of representatives from listed companies and stakeholders shows that the tendency and direction of sustainability communication strategies are separated into internal and external perspectives that influence the company. From the internal perspective, listed companies are assumed to integrate participation by the company and each group of stakeholders towards sustainability. Tangible and intangible results from sustainability must be considered and analysed for appropriateness to stakeholders. While strategic communications must be implemented for all sustainability strategies, communication tactics and channels must be integrated. This is done with information that must be easy to access, explicit, and transparent. Moreover, listed companies should implement crisis communications and risk management to enhance continuous business management.

The external aspects that influence the sustainability communication of listed companies are expectations, needs, suggestions, and comments from all stakeholders who are related to the organisation to design content that meets their expectations and needs. Communication channels and tactics are then selected, especially digital media such as websites and social media, and the language is modified to suit each group of stakeholders. In addition, listed companies should follow achievement goals for a unified society. For more practical results, compliance with sustainability policy set by the government is necessary. The policy could be in the form of a request for cooperation or the issuance of rules and regulations.

In terms of the tendency and direction of sustainability communication, "stakeholders" are the target audiences that all listed companies are paying attention to. Many acquisitions and executions, including information acquisition, content identification, communication tactics and channels, should be made appropriate for each group of stakeholders. On the issue of content development, the stakeholder groups in Thailand believe that listed companies should focus on sustainability issues that affect the economic, social, and environmental dimensions. This in consistent with the research titled *Sustainability Management with Traditional and Innovative Social Media Communication Tools*, by Toppinen and Hanninen (2013), that was produced during the XXIV (24th) ISPIM Conference with the theme Innovating in Global Markets: Challenges for Sustainable Growth in Helsinki, Finland. The study was conducted with 100 sustainable companies in the paper industry. The environmental issue was a matter that the senior management of the organisations had emphasised and communicated to the stakeholders. Social media were focused on social issues instead. DeMers (2013) also offered additional online content, besides providing corporate sustainability information, including answering questions from audiences, communicating general information, posting job announcements, and carrying out customer satisfaction surveys regarding needs and opinions of products or services. Communication challenges for today's companies include the means of communication to understand the issues among target audiences that affect business survival such as preserving natural resources, creating supply chains, as well as promoting efficient energy usage.

Moreover, as to the matter of content, Villarino and Font (2015) explored sustainability communication in the hospitality business. The researchers found that the message and content of sustainability communication were more about the presentation of facts than emotional content and were not presented as specific to each customer. Even though the communicated messages were clearly understood, they were passive rather than encouraging participation. They have a positive effect on perceptions but do not create an impressive image of the business.

Adjusting the format of the message for communication as well as customer focus cannot be overlooked. Listed companies should develop communications for persuasion. The persuasive content must comprise four dimensions: (1) the characters and groups of the target audience, in order to produce appropriate messages; (2) the substance to be structured in such a way that creates persuasiveness, such as direct and indirect content, as well as symbolic and clear content; (3) the consideration of persuasive content that draws attention, is rational, and follows the norm, as well as content that reflects the experience of the target audience; and (4) the use of data from trusted and certified organisations as significant information sources to build credibility.

On the issue of communication tactics, digital media are highly influential communication channels. This outcome is in line with Dovleac's (2015) research that examined the growth of online media, especially websites, Twitter, and Facebook, that contribute to the development and transformation of human behaviour and interactions, especially in terms of awareness and transmission of sustainability issues in both their social and environmental dimensions. Online media also play an important role in enabling individuals to communicate and share knowledge and opinions with others (Budden et al., 2011; Kumar et al., 2010). This is also consistent with the *USA Today* survey that found that more than two-thirds of large organisations use social media and blogs to communicate (Mullaney, 2012). In addition, more than 66% of the world's largest, Fortune 500-rated organisations are still conducting corporate communications via Facebook channels and corporate tweets. Companies and small businesses also use these types of media as they are low in cost and able to reach recipients around the world (Heggestuen & Danova, 2013). Factors that influence companies to focus on online communication from Dovleac's (2015) study include the following issues: (1) online communications make it easier for organisations or companies to engage with customers (Mangold & Faulds, 2009; Tiago & Verissimo, 2014); (2) consumers trust other individuals within their online communities rather than believing the companies (Kotler et al., 2010); (3) the size of the online population has increased dramatically, and, thus, businesses need to improve and develop their operation systems (Ngai et al., 2015).

In addition, Scheel et al. (2015) have also presented an additional perspective on the influence of online media: (4) companies must shift from offering products and services to connecting and interacting with customers; (5) organisations must be streamlined to meet the needs of consumers; (6) information must be transparent; and (7) data access requires easier internet connections; and (8) many organisations use online media to maintain relationships with existing groups and build relationships with prospective customers. Even sustainability reports and social responsibility reports are being presented online (SMI, 2012). The focus on stakeholders is consistent with stakeholders' perspectives, as strategic communications is a strategy that communicates in several dimensions, such as economic, social, environmental, cultural, and political, with the stakeholder groups to which the organisation is relevant, with appropriate content

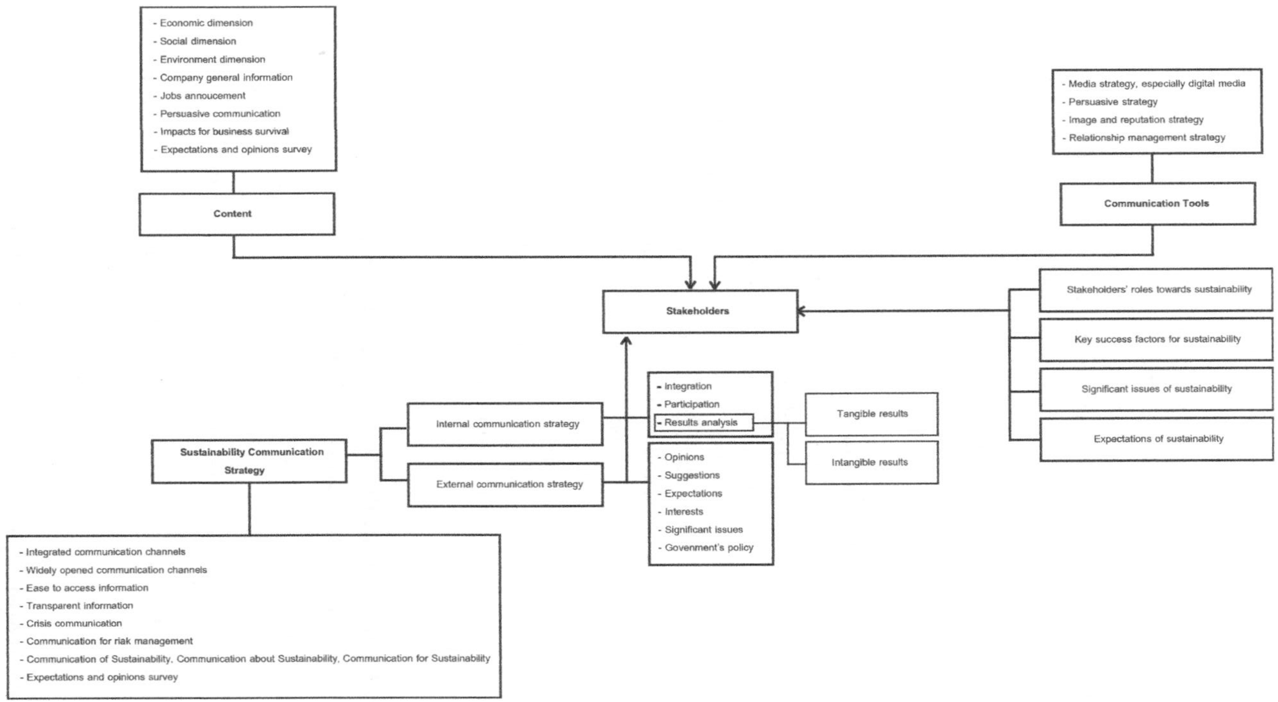

FIGURE 4.4 The tendency and direction of strategic sustainability communication among companies listed on the SET

and communication channels. It aims to build awareness, understanding, and participation and change attitudes or behaviours.

The use of online communications, especially on websites, has led to Lodhia's (2014) research that proposes websites' contribution to promoting social and environmental issues, also known as sustainability communication. Users should take into account the five factors that affect the use of websites for sustainability communication, namely: (1) economic factors, whereby users have to consider the media costs of preparation and maintenance throughout the usage period; (2) internal factors, identifying a clear communication strategy for sustainability; (3) the stakeholder groups as a factor outside the organisation – the company needs to acknowledge the needs and expectations of the stakeholders to communicate appropriately; (4) the potential impact of a double-edged sword in that the content presented on the website is tailored to the needs of one group and may not meet the needs of another group; it may also present information that may affect other stakeholders and place pressure on the company; and (5) corporate restructuring after acquisition by global companies may also result in similar communication content and strategy of the company. This may lead to communications altered in content and substance from the local. Stakeholders of the same social group may claim information that is consistent with local operations but not internationally. The study also highlights the effect of using websites to present more in-depth concepts than just presenting and reporting to gain insights into sustainability. Moreover, this study also discusses social and environmental issues with in-depth insights into stakeholders' needs that must be communicated on the websites.

The issues of sustainability communication strategies, content, and tactics based on the interviews of representatives and relevant research related to sustainability communication have been analysed, and the tendency and direction of the sustainability communication of companies listed on the SET are illustrated in Figure 4.4.

The Strategic Sustainability Communication Model of Companies Listed on the SET

From the exploration of the implementation of sustainability communication strategies by companies listed on the SET, the researcher proposes the similarities and differences in strategic sustainability and strategic sustainability communication as shown in Table 4.2.

From Table 4.2, the researcher has considered the results and issues from the perspectives of the listed companies' representatives and stakeholders, as well as international studies, in order to develop broader viewpoints of the strategic sustainability communication model in SET. The nine processes of the strategic communications model from Mahoney (2009) are also compared with the similarities and differences. Finally, the strategic sustainability communication model of companies listed on the Stock Exchange of Thailand is illustrated in Figure 4.5.

TABLE 4.2 The similarities and differences in strategic sustainability and strategic sustainability communication of companies listed on the SET

Topic	*Similarity*	*Difference*
Strategic sustainability implementation	Three dimensions of strategic sustainability framework: • economic • social • Environment	• Corporate governance dimension • Risk management dimension • To develop self-framework
Strategic sustainability process	• Situation analysis • Vision identification • Indicators and goals identification • Budget and action plans approval • Action plans implementation • Review and evaluation	
SET's strategic sustainability guidelines	• To comply with SET's sustainability guidelines • To be member of Dow Jones Sustainability Index and comply with guidelines	
Factors of strategic sustainability implementation	• Internal factors • External factors	
Strategic sustainability communication	• Stakeholders classification • Stakeholders' expectation analysis	
Type of sustainability communication	Appropriate communication channel	
Tendency and direction of strategic sustainability communication	• To focus on stakeholders as target audience • Online media • Appropriate communication channel selection • Appropriate contents development	• Information transparency and convenient to access • Integration • Compliance with government regulations • Crisis communication management • ROSI analysis

Conclusion

Companies listed on the Stock Exchange of Thailand should consider the economic, social, and environment dimensions or develop these three dimensions to build a sustainability strategic framework. Strategic sustainability should consist of six processes, which are situation analysis, vision identification, indicators and goals identification, budget and action plan approval, action plan implementation, as well as

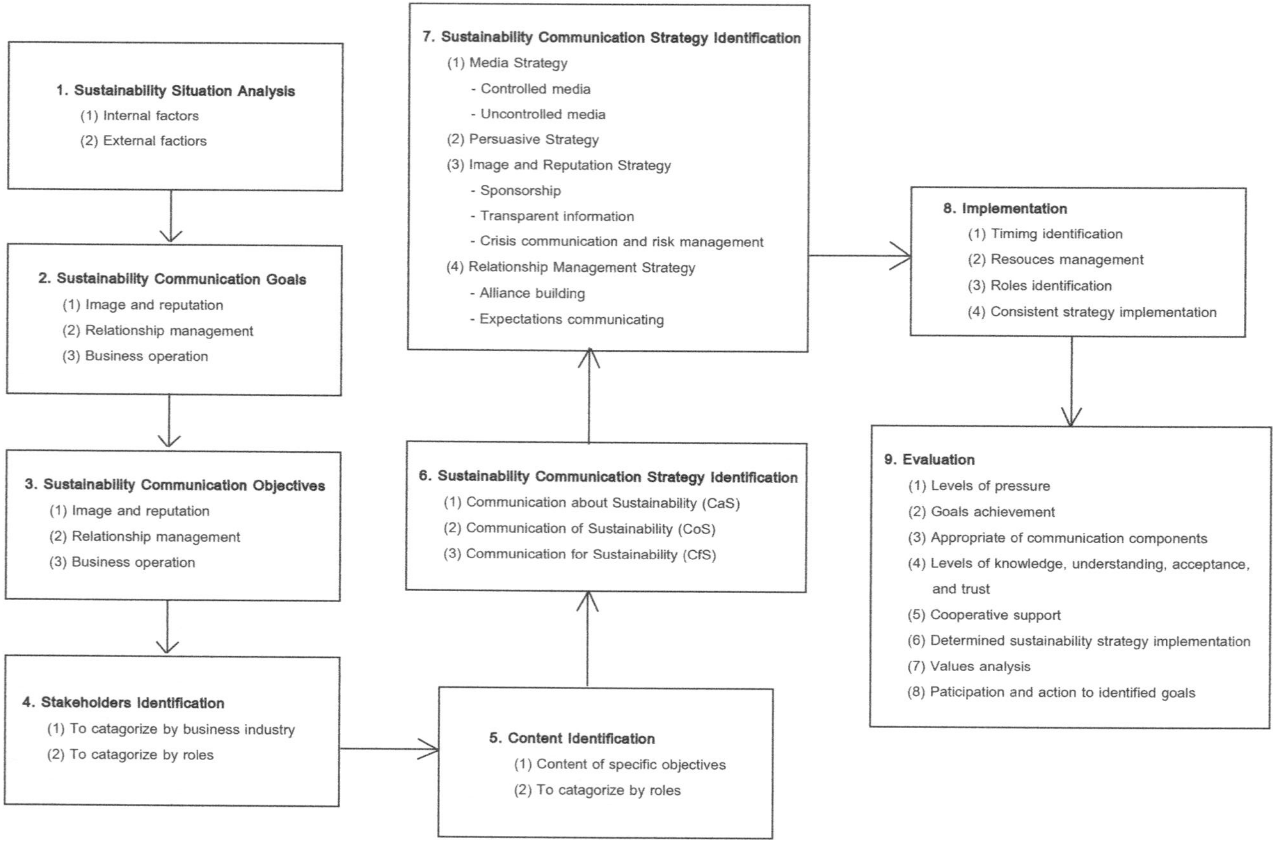

FIGURE 4.5 Strategic sustainability communication model of companies listed on the SET.

review and evaluation. Moreover, there are three issues that listed companies should be concerned with, which are (1) timing, as long-term vision identification; (2) existence, by improving and developing compatibility with each period of time; and (3) stakeholders, by exploring their expectations and serving them appropriately.

Listed companies should implement strategic sustainability communication by exploring and serving stakeholders' expectations. Strategic sustainability communications are identified by nine processes based on the following procedures: (1) sustainability situation analysis; (2) sustainability goals identification; (3) objectives and indicators identification; (4) stakeholders identification; (5) content identification as facts and persuasive format; (6) serving stakeholders' expectations; (7) communication tactics identification; (8) strategies and tactics implementation; and (9) communication evaluation in terms of perception, awareness, understanding, and participation, as well as the consideration of attitude and behaviour change.

The stakeholders from listed companies should significantly concentrate on "employees". This is because "employees" will be the communicators who enhance understanding of sustainability issues by external stakeholders depending on the departments and sectors that they are working with. For the other groups of stakeholders, listed companies should understand and realise roles and prioritise issues in order to serve their expectations and needs appropriately.

Listed companies should utilise digital media such as websites, Facebook, and Twitter as the preferred channels on which to broadcast information to audiences, deliver content widely, edit instantaneously, and sustain communication. Though beneficial, the negative effect of digital media is that they may ruin the corporate image of a company if an issue raised by stakeholders is not addressed promptly. Therefore, listed companies should initiate a crisis communication strategy and business continuity management by involving the communication department.

Listed companies must understand the characteristics and differences of these sustainability communication strategies, which are communication of sustainability (CoS), communication about sustainability (CaS), and communication for sustainability (CfS), in order to implement tactics that serve the intended communication objectives and goals. This is to enhance cooperative sustainable development between companies and stakeholders, timely transformed attitudes or behaviours, and the integration of all sustainability communication strategies.

There should be continuous integrated cooperation between the government, private sectors, the Stock Exchange of Thailand, and civil society to drive sustainability by formulating principles, complying with rules and regulations, communicating for understanding, participating, as well as changing attitudes and behaviours. In enhancing the holistic effectiveness of internal and external communications, close cooperation between the strategic sustainability and corporate communications department must be in place. Issues that must be continuously prioritised in sustainability communication by listed companies are stakeholders' expectations, credible information and facts, as well as the presentation of new forms of data to adapt to the rapidly changing media and dynamic social contexts.

Bibliography

Argenti, P., Howell, R.A., & Beck, K.A. (2005), The strategic communication imperative. *MIT Sloan Management Review, 46*(3), 83–9.

Barth, M. (2012). Social learning instead of educating each other. *GAIA, 21*(2), 91–94.

Bonini, S. (2012). *The business of sustainability*. Chicago: McKinsey & Company.

Cornelissen, J.P. (2005). *Corporate communications: Theory and practice*. London: SAGE.

Drost, E. (2011). Validity and reliability in social science research. *Education Research and Perpectives, 38*(1), 105–123.

Friedman, T. L. (2009). *Hot, flat, and crowed: Why we need a green revolution-And how it can renew America, release 2.0*. New York: Picador/Farrar, Straus, and Giroux.

Funtowicz, S., & Ravetz, J. (1993). Science for the post-normal age. *Futures, 25*(7), 739–55.

Godemann, J., & Michelsen, G. (2011). Sustainability Communication: Interdisciplinary perspectives and theoretical foundations. In *Sustainability communication* (pp. 3–11). Dordrecht: Springer.

Grossman, N., Algie, J., Treerutkuarkul, A., & Wegner, N. (2015). *Thailand's sustainable development sourcebook: Issues and information, ideas, and inspiration*. Singapore: Editions Didier Millet.

Grunig, J.E. (2006). Furnishing the Edifice: Ongoing research on public relations as a strategic management function. *Journal of Public Relations Research, 18*(2), 151–76.

Hallahan, K., Holtzhausen, D., Van Ruler, B., Vercic D., & Siramesh, K. (2007). Defining strategic communication. *International Journal of Strategic Communication, 1*(1), 3–35.

Khapa, L. (2016). *The 20-year national strategy, sustainable development goals and the 12th national economic and social development plan of Thailand*. Accessed from http://www.tsri.or.th

Luhmann, N. (1995). *Social system*. Stanford: Stanford University Press.

Mahoney, J. (2009). *Strategic communication: Principles and practice*. Australia: Oxford University Press.

Moser, S. C. (2010). Communicating climate change: History, challenges, process and future directions. *WIREs Climate Change, 1*(1), 31–53.

National Strategic Plan Committee. (2016). *The 20-year national strategy (2017–2037)*. Accessed from https://www.royalthaipolice.go.th

Nerlich, B., Koteyko, N., & Brown, B. (2010). Theory and language of climate change communication. *WIREs Climate Change, 1*, 97–110.

Newig, J., Schulz, D., Fischer, D., Hetze, K., Laws, N., Lüdecke, G., & Rieckmann, M. (2013). Communication regarding sustainability: Conceptual perspectives and exploration of societal subsystem. *Sustainability, 5*, 2976–2990.

Nidumolu, R., Prahalad, C. K., & Rangaswami, M. R. (2009). Why sustainability is now the key driver of innovation. *Harvard Business Review, 87*(9), 56–64.

Reilly, A., & Hynan K. (2014). Corporate communication, sustainability, and social media: It's not easy (really) being green. *Business Horizon, 57*, 747–758.

Sandhu, S. (2009). Strategic communication: An institutional perspective. *International Journal of Communication, 3*(2), 431–46.

Tewari, R., & Dave, D. (2012), Corporate social responsibility: Communication through sustainability reports by Indian and multinational companies. *Global Business Review, 13*(3), 393–405.

The Stock Exchange of Thailand. (2015). *SET awards*. Accessed from http://www.set.or.th/th/news/issuer_activities/setawards/setawards_p1.html

United Nations. *Transforming our world: The 2030 agenda for sustainable development.* Accessed from http://sdgs.un.org

Van Marrewijk, M., & Werre, M. (2003). Multiple levels of corporate sustainability. *Journal of Business Ethics, 44*(2/3), 107–119.

Ziemann, A. (2011). *Communication Theory and Sustainability Discourse.* Dordrecht, The Netherlands, New York: Springer.

5

PERSUASION IN SUSTAINABILITY COMMUNICATION

A Study of Penang Green Council's Initiatives

Debbita Tan Ai Lin and Shaidatul Akma Adi Kasuma

Introduction: Sustainability Communication

Effective communication involves a two-way process for the transfer of ideas, thoughts, and feelings via verbal or non-verbal means in a simple, direct, and precise manner (Genç, 2017). Communication is key to promoting sustainability, defined as the act of meeting the needs of existing citizens without compromising the ability of future generations to meet their needs and the "long-term and difficult goal of reaching an ecologically sustainable state" (Dovers & Handmer, 1993, p. 217). "Sustainability" comes from the verb "to sustain", referring to the possibility of continuing over a long period of time.

Sustainability communication is the act of communicating sustainability messages that include environmental issues, social justice, and the economy (Mohamad Saleh, 2017). It aims to establish societal transformation on the basis of normative goals of sustainable development and is measured by its impact (Genç, 2017). In an organisation, sustainability communication is an influential corporate strategy to demonstrate an organisation's intention to help eradicate ecological concerns, enhancing their reputation in a way that may contribute to income generation. Corporations are trusted entities, and their advocacy of a green lifestyle has a huge impact on consumers' choice (Kapoor et al., 2021).

The emergence of new media and a large global audience base encourage organisations to communicate their sustainability endeavours online. New media include a diverse range of platforms, from newspaper articles, blogs, and websites to social media, music, podcasts, and forums, transmitted in digital form. Social media channels, particularly Instagram and Twitter, have been identified as broadening and supporting green conversations between organisation and consumer and consumer and consumer (Milanesi et al., 2021; Saran & Shokouhyar, 2021; Yeniçeri Alemdar & Kocaömer, 2020). Articles and news releases disseminated

DOI: 10.4324/9781003309642-5

via social media are more effective as persuasive consumer–consumer messages that stimulate eco-purchasing than advertising (Byrum, 2019). Work-related content posted on social media is considered a credible source of information and affects the way employers and customers perceive an organisation (Pekkala & van Zoonen, 2022). Therefore, a good organisation–consumer relationship reflects impressive corporate social responsibility initiatives and leads to impactful sustainability messages (Kapoor et al., 2021).

The Role of Media in Green Communication

Communication technology and mass media largely influence public opinion, promote ideology, and shape sentiment. They act as an accountability mechanism in raising societal awareness of issues that may otherwise never get publicly debated or addressed. For example, news about controversial projects such as rare earths mining in Malaysia may stimulate social policy and a change in the curriculum to produce civic competence (Rossi, 1996). Civic competence is formed via thoughtful social interaction among the public (Rossi, 1996), and, thus, the media's exposure of an issue directly impacts the development of human capital.

Consequently, trusted, reliable media sources are paramount in the propagation of technical and scientific issues in a contemporary risk society with accelerated social and digital transformation (Anderson et al., 2012). The media may also become an instrument against dishonest, fabricated, and biased arguments that aggravate situations and divide communities. In the age of new media that give voice to everyone with an internet connection, the media must further exercise their role in channelling truthful, impartial data to create an informed and conscious society. New media allow individuals or communities to articulate their thoughts, ideas, and opinions freely in online spaces created for like-minded people. This practice gives rise to diverse communities, one being the community of sustainability advocates. In German newspaper media, the youth climate change activist Greta Thunberg was described using derogatory, ageist language, such as "pupils", "absentees", and "dreamers", to invalidate her ecological movement (Bergmann & Ossewaarde, 2020).

As a universal concern, sustainability discourses are increasingly integrated into the public domain as well as mainstream, traditional, and new media (Abdullah et al., 2009). Naturally, firms that have been practising green communication via social media were more active than ones that were not green in addressing sustainability issues, as well as general activities. Depending on the nature of firms, the sustainability initiatives, communication media, and metrics employed vary (Reilly & Hynan, 2014). Quality, continuous communication via multiple channels (e.g. speeches, personal communications, press releases, web pages, manuals, newsletters, articles, videos, and social media channels) has the ability to motivate desired sustainability action. For instance, Luu (2021) found a significant relationship between quality green communication and employees' intention to reduce food waste at the workplace and at home. This observation further emphasises the central role media play in cultivating and popularising a sustainable lifestyle.

Differences may also be seen between organisations with different purposes and types of business. Based on the findings of individual words on websites, environmental non-governmental organisations (ENGOs) and green businesses have unique conceptualisations of sustainability. ENGOs described in detail the natural environment and their perception of damage and responsibility, while green businesses centralised good intentions and risk management (Brown, 2013). Many organisations are in fact occupied with "sustainability correctness" in aligning themselves with a dogmatic inclination. They are being coerced into using the "right" sustainability words instead of conveying alternative perspectives that might be radical and drastic (Stibbe, 2009). In this regard, Joosse and Brydges (2018) found personal green blogs interesting to study as they showcased individuals' political and cultural beliefs and practices on sustainability issues that may differ radically from those of the conventional media in legitimacy, form, and content. Therefore, the emergence of new media is timely and empowers the community of eco-consumers to converse about a wider range of ecological matters.

In Malaysia, the media and ENGOs act as informers, educators, and researchers on environmental and sustainability issues (Mohamad Saleh, 2017). The Malaysian media, however, do not have the privilege of total freedom when reporting. While given an extent of autonomy to express their ideology, many media personnel articulated less liberty in retrieving environmental information, especially that associated with renowned corporations and scandalous undertakings (Mohamad Saleh et al., 2018). Yacoumis (2018) similarly observed Australian newspapers' avoidance of non-hegemonic ecological and social discourses, which diminishes the public's desire to engage in sustainability discourses. On top of that, Lowrey et al. (2007) posited journalists' insufficient preparedness and insufficient expertise in science and medicine training to cover communication of disasters that include terrorism, as well as natural and human-initiated adversities. With the advent of social media, everybody can publicly disseminate information, and people without proper training often breach proper journalism etiquette.

The Language of Ecological Discourse on New Media

Sustainability communication is highly dependent on the content, language, and visuals used for engagement. A well-crafted sustainability message with supporting visuals has the power to influence the audience's decision and alter behaviour.

Persuasive messaging techniques are key in any sustainability communication initiative. Effective eco-messages (content) are carefully constructed at the grammatical, lexis, and syntax level. Green (2006) and Luu (2021) found narratives to be an effective means of transforming behaviours into more sustainable ones. Stories reduce counterarguments, facilitate mental simulation of the unknown, provide a model for behaviour change, and create strong attitudes based on cognition and emotion (Green, 2006). Audiences who are immersed in narratives are persuaded to change their beliefs that motivate actions.

Nwagbara and Belal (2019) reported persuasive language at three levels; single word (wording), phrase, and chain of words. In the single word strategy, nouns, pronouns, verbs, adjectives, and adverbs are used to convey a sense of a responsible organisation and its relations. These aspects are also nominalised to demonstrate semantic meaning of sustainable development. At the phrasal level, indivisible parts of grammatically accurate clauses make up sentences. In the corporate social responsibility (CSR) reports of Nigerian oil companies, the persuasive phrases are constructed with "purposive so and to" – for example, "To further improve the RDC's effectiveness" (p. 2418). In the chain of words strategy, the subject of a sentence (e.g. noun, noun phrase, pronoun) relies on a (subordinate) conjunction to have logic and validation in legitimising action. An example of a word chain strategy is in the sentence, "We will continue to operate with the highest standards of safety, environmental performance, ethics" (p. 2419).

Relatedly, Grabowski (2007) found several persuasive techniques in the discourses of financial providers based on Fairclough's critical discourse analysis (CDA). These include the "problem–solution" relation (e.g. "If you need money, we are here to give you a loan"); grammatical manipulation (repetition of a positive imperative clause, e.g. "Apply now", "Call now"); semantic manipulation (to make assumptions and presuppositions, e.g. "Upgrade to Platinum and 2000 reward points are waiting for you"); synthetic personalisation and conversationalisation (the use of personal pronouns such as "you" and "we"; e.g. "Tell us how much cash you need"; pp. 4–6).

When discussing climate change, US and Spanish media all used similar hedging words, though a US newspaper employed a higher density of epistemic markers and more ambiguous grammatical constructs of uncertainty (Bailey, Giangola & Boykoff, 2014). The frequency of the hedges remained the same or increased with time, despite the relative scientific consensus that human activities cause climate change. The US newspaper frequently highlighted differences between the Intergovernmental Panel on Climate Change (IPCC) and scientific observations to raise uncertainty. This study is a good example of how the media control information that shapes ideology. In alleviating ambiguity and increasing transparency in the fashion world (Hussain, 2017), brands and retailers used a combination of visual data, text, report, and design to communicate the sustainability of their supply chain via e-commerce and corporate fashion websites. Improved transparency enhances social accountability; this translates into information accessibility and quality, openness, and effective design (perceived transparency), which is the very tenet of new media (e.g. websites and social media).

On Twitter, members of the public have published more green tweets than corporations, substantiating eco-awareness on social media (Saran & Shokouhyar, 2021). Furthermore, organisations' green contents received higher retweets compared with non-green sharing. In this regard, sustainability communication is fundamentally a public relations strategy to keep the stakeholders in an interaction loop and influences the public's perceptions of, and behaviours in relation to, a firm's ecological intention. It is a conscious effort to pacify the growing community of eco-consumers and establish a firm's stand as a responsible, resourceful, and observant entity (Jha & Verma, 2022).

Eco-consumers reciprocate positively to companies that demonstrate high sustainability communication activity, such as engaging in frequent communication on social media, by being loyal, and placing more trust in their endeavours, resulting in a higher revenue (Jha & Verma, 2022; Smith, 2021). A higher level of retweeting/sharing of posts indicates a higher interaction and a stronger sustainability communication and consumerism (Saran & Shokouhyar, 2021). Smaller companies with higher interactions tend to value eco-labelling in their online offers and seem to benefit more compared with the big corporations, which amplifies the opportunities leveraged by social media in eco-consumer markets (Jha & Verma, 2022; Tiago et al., 2021).

Instagram's forte as an image-sharing platform is linked to its ability to generate two-way interactions – that is, dialogue, interpersonal communications, and digital engagements – in promoting environmental causes, such as sustainable fashion and eco-wear (Milanesi et al., 2021; Yeniçeri Alemdar & Kocaömer, 2020). This is achieved by using selfie images and videos and first-person pronoun-rich captions (e.g. "I", "we", "us"), and responding to comments (Martin & MacDonald, 2020). In this instance, Instagram offers a greater basis for dialogic communications than Twitter. Social media ease the process of audience-to-audience communications and spread news releases and articles detailing CSR efforts to stimulate eco-purchases (Byrum, 2019). The use of interpersonal communication strategies encourages conversational engagement, shown by the number of comments and unique individuals involved in conversations (Martin & MacDonald, 2020).

On the other hand, ineffective sustainability communication leads to the loss of income. Examples drawn from the tourism industry show that unsuccessful ecological messages led to low bookings and non-appreciation of products (Tölkes, 2020). This was because, despite their initial interest in sustainability packages, tourists faced difficulty realising, recognising, and understanding the sustainability attributes of the products they had booked. To overcome this, Jha and Verma (2022) call for more digestible, comprehensible nuggets of information often observed in tweets on Twitter. Besides that, the consumer communicator role is viewed as significant in the growing trend of sustainable lifestyles.

Persuasion and Sustainability Communication

Generally speaking, sustainability communication represents sets of rhetorical codes that are not solely meant for showcasing green efforts by a particular organisation or entity but also to convince the audience to participate in the very same initiatives. It is, in essence, about agency. Just as political communication can be construed to be driven by desires to gain support from the masses and to legitimise power or governance, sustainability communication can be, and rightfully should be, about encouraging communities to come forward to realise green initiatives.

Manipulative or consensual, persuasion is persuasion, and the promotional culture has consistently permeated society across time. Persuasion within the aspect of altruism is reciprocative in some sense; the targeted audience satisfies the persuading entity's need for action while enjoying the emotional rewards of giving. It is a social exchange. People do experience satisfaction when they choose to participate in a cause that they perceive to be good. In a study on participation in green programs, Giebelhausen et al. (2016) found that "people are more satisfied with a service experience when they choose to participate in the provider's voluntary green program (e.g., recycling) – an effect mediated by the 'warm glow' of participation" (p. 56).

Incentivisation (compensating the programme participants) paradoxically increased satisfaction for those who did not participate, but decreased satisfaction among those who did. Giebelhausen et al. (2016) suggested the use of mixed incentive bundles (providing both self-benefiting and other-benefiting options) to maximise the "warm glow" effect and satisfaction for both groups.

Sustainability communication is, on the whole, encapsulated within the cocoon of altruism, and, while there are various key elements to consider, rhetoric remains central. In other words, the potential emotional rewards of giving alone are not, generally speaking, sufficient for agency. We opine that the art of effective or persuasive communication via writing or speaking plays a more crucial role, in tandem with the potential emotional rewards of giving, for any agenda geared toward the eventual realisation of green initiatives.

In their work on organised persuasive communication, Bakir et al. (2019) observed that research in the fields of propaganda and persuasion has grown exponentially since the early 20th century, noting also the advent of new internet-based media environments. These new, digital arenas are powerful media through which ideas can be communicated and actions induced. Conditions and consequences arising from social and environmental phenomena are exigencies that render the study of green rhetoric a critical undertaking. Healing the planet generally does not feature very prominently in our minds, despite news of unsettling environmental effects that punctuate the media regularly. In 2021, Reuters reported on the UN's (United Nations) climate change report which was described by UN Secretary-General António Guterres as a "code red for humanity" (Chestney & Januta, 2021).

Largely, people are quite apathetic in terms of not being "eco-righteous". Different factors are often at play, of course, but scholars of rhetoric would have us view rhetorical analyses of sustainability discourse (or "salesmanship", to borrow a term from Bakir and colleagues, 2019) as more prescriptive than normative, because an extended understanding of the rhetorical strategies used to coax and to provoke better treatment of the environment can only benefit all strata of society – of the present generation and beyond.

Aristotelian Framing

Persuasion has always held court at the heart of any salesmanship. The extent of successful persuasion is essentially correlated with how well one appeals to the audience. The Greek philosopher Aristotle (384–322 BCE) taught that one can appeal in different areas, primarily in the areas of *logos* (logic), *ethos* (credibility), and *pathos* (emotion). Considered in partnership, these appeals form what later rhetoricians term the "rhetorical triangle".

This persuasion triad is a multifaceted doctrine, one that permits us to classify and discuss threads of communication derived from diverse sources and events across time. What Aristotle contributed to rhetorical theory was fundamentally the establishment of how persuasion should be treated.

To Aristotle, a speech mainly involves three things: the speaker, the subject of the speech, and the listener. Therefore, these are paramount: whether the speaker is worthy of credence, the argument itself, and the emotion(s) of the listener. For the first, it matters even if the speaker just appears credible. More importantly, the audience must agree that there is indeed a problem that needs to be resolved.

The heuristics embodied in Aristotelian appeals have been employed variably over time to deliver persuasive discourse and even designs, be it in the name of politics, commercial marketing, or public service. For instance, Tan and Al-Rikaby (2022) analysed a 2021 televised interview, an exclusive on the COVID-19 pandemic, involving then Malaysian Prime Minister Muhyiddin Yassin. The analysis was based on the Aristotelian persuasion triad and also included the lesser known appeals of *telos* (purpose) and *kairos* (timeliness). Overall, the use of *logos* and *ethos* dominated the interview, with *pathos* only insignificantly manifested.

On the green front, Voci (2022), in her work *Logos, Ethos, Pathos, Sustainabilitos? About the Role of Media Companies in Reaching Sustainable Development*, focused on the two largest commercial media companies in Germany and highlighted that most sustainability-related content addressed food issues, followed by issues concerning resources and the environment, in line with some of the 17 Sustainable Development Goals (SDGs). Furthermore, these issues were primarily communicated logically, appealing to the audience's reason. The ethical *ethos* did not feature prominently, and the emotional *pathos* scarcely occurred.

Framing is essential for any public narrative aimed at convincing others to act. Framing in communications science refers to how an issue is portrayed and understood (Supran & Oreskes, 2021); analysing frames allows us to discern how actors have attempted to shape discourse in order to achieve certain goals. A potent framing of responsibility, for example, can be a determining factor in decisions to endorse war and violence, or the reason why people begin to significantly reduce their carbon footprint.

According to Allen (2017), *logos* is one of the three proofs or appeals and provides a basis for reasoning and argument. That which is logical is appealing because it makes sense and is acceptable. The second appeal described by Allen

TABLE 5.1 Aristotelian framing in communication

Rhetorical Appeal	*Definition*	*Indicator*
Logos	Appeal to *logic*	Use of reason, facts, evidence, and statistics to support a statement or argument
Ethos	Appeal to *credibility*	Establishing trust by demonstrating knowledge, expertise, abilities, and ethical behaviour, which includes acknowledging the rights and feelings of the target group Use of ingratiation and self-deprecation
Pathos	Appeal to *emotion*	Use of emotive expressions, sensory images, and anecdotes to manipulate feelings

is *ethos* – persuasion through knowledge and character. It is natural for an audience to be persuaded by someone or an entity they believe to be credible in these aspects. Additionally, Aristotle himself recognised the special role of *ethos* in communication, observing that it may almost be deemed the most powerful means of persuasion at a proponent's disposal (Koszowy et al., 2022). For *pathos*, Allen (2017) pointed out that it is inherently emotional manipulation, and Aristotle defined emotions as states of mind involving both pleasure and pain, which can in turn affect our perceptions and actions. Table 5.1 elucidates the three appeals, extracted from Tan and Al-Rikaby (2022).

Humanised Discourse, Greek Thinkers, and the Malaysian Landscape

Sustainability communication and, more specifically, green communication are hardly new phenomena. Godemann and Michelsen (2011) explained that the development of the term "sustainability communication" occurred in tandem with the call for humankind's responsible interaction with its social and natural environments. They also brought the following to the fore:

> This entails a process of social understanding that deals with the causes and with possible solutions. The task of Sustainability Communication is to critically evaluate and introduce an understanding of the human–environment relationship into social discourse.
>
> *(p. 3)*

It is fair to suggest, therefore, that, in order to motivate society to make more pro-sustainable choices, sustainability discourse must be humanised. It must demonstrate the congruence between what is problematic and what people *can* do to make things better, and also (ideally) to prevent more problems from arising.

Abdul Rahman (2020) mentioned that "sustainability" has been a buzzword since the 20th century, pointing out that its history is much older. This concurs with Cooper's (2001) opinion of Aristotle, whom he saw as an environmental thinker:

> Aristotle was not, of course, an environmental scientist or philosopher in the contemporary sense ... [since] the "eco-crises" which have stimulated recent environmental concerns were happily unknown in ancient Greece ... it is clear however that Aristotle experienced and urged a profound regard for the living world and several elements of his thinking prove attractive to contemporary environmental thought.
>
> *(p. 12)*

Ancient Greek philosophers played their part in providing the foundations for environmentalism and environmental ethics. Even though the general view was that human intervention was not necessary to preserve nature, classical Greek philosophy stressed a deep respect for the intrinsic value of nature. Westra and Robinson (1997) noted that environmental ethicists have frequently criticised ancient Greek philosophy as anti-environmental, partly because of a view of the world that puts nature at the disposal of humans. However, they also noted that some viewed the Greek thinkers as having nevertheless furnished us with an excellent foundation for a sound theory of environmentalism.

Although sustainable communication in the Western sphere reaches further back in history and has gained a steady foothold in contemporary mainstream media, the same cannot be said in the Malaysian context. Considering Malaysia's rather substantial environmental problems, one would expect the local media and environment-oriented organisations to be much more spirited in communicating sustainability messages to the *rakyat* (Malay for Malaysian people), not solely to increase awareness but also to establish more pro-environmental behaviours. Yet, sustainability issues appear to be less favoured in comparison with other news, such as news concerning political antics. Mohamad Saleh (2017) elaborated that, sadly, environmental news is usually placed together with other news types (specifically under science and technology), and that Malaysian research covering environmental sustainability communication is limited.

In a 2018 publication, Mohamad Saleh, Heinrichs, and Nik Hasan focused on the struggles of the Malaysian media and ENGOs in communicating environmental issues to the public within a semi-democratic nation, stating that the media and ENGOs in Malaysia do not actually experience true liberty in terms of communicating environmental issues to the masses.

Malaysia's efforts to be in line with sustainability and environmental concerns can be seen in its advocacy for Agenda 21, which promotes environmental education and awareness among the people (Mohamad Saleh et al., 2018). Agenda 21, a world action plan resulting from the United Nations Conference on Environment and Development held in Brazil in 1992, is fundamentally an agenda to achieve sustainable development in the 21st century.

The role of the media in educating the public about the environment is included in the Stockholm Declaration on the Environment; it comprises a proclamation of 26 principles and submission of recommendations. Principle 12 states:

> Resources should be made available to preserve and improve the environment. Education in the environmental matter is essential and mass media should help in this.
>
> *(More, 2019)*

Within the Malaysian context, the key roles of the media and ENGOs in this regard have been specified by the government under the Seventh Malaysia Plan and the Economic Planning Unit (Mohamad Saleh et al., 2018). However, the media found it difficult to focus on their roles as environmental educators as they had to cover other news as well, whereas ENGOs confirmed the problem of limited funding and highlighted that Malaysians are not interested in learning about issues pertaining to the environment (Mohamad Saleh et al., 2018).

More recently, Mohamad Saleh and Heinrichs (2020), in a paper on media representation of environmental issues in Malaysia, revealed that the media have given more coverage to environmental problems such as deforestation and pollution than to green efforts such as environmental conservation and education, because problems are more dramatic and attention-grabbing than efforts. In contrast, ENGOs accorded more attention to effort topics such as sustainable living.

Sustainability Communication: The Case of Penang Green Council

The present study is an illustrative example of how language can be employed as an agent of persuasion within the context of green discourse. We, the authors, prefer to stick with caution and not make definitive claims. We do, however, find it reasonable to state that persuasion is often, if not always, inherent in public discourse. It is also possible for such discourse to be interpreted beyond its intended meanings. Analysing a particular corpus based on a logical framework generally permits us to draw a number of conclusions reasonably. We aimed to study Penang Green Council's website discourse (https://pgc.com.my/2020/), with particular focus on its green initiatives (Beat Plastic Pollution and Waste Segregation at Source). Analysis was carried out based on the Aristotelian constructs of *logos*/logic, *ethos*/credibility, and *pathos*/emotion.

Established more than a decade ago, the council is supposed to nurture, facilitate, and coordinate environmental causes for a cleaner, greener, and safer Penang. The website also mentions that the state government mandates investments in vigilant green efforts.

Logos

Upon accessing the Penang Green Council (PGC) website, one is first exposed to an introductory page with the following statements:

Extract L1

> Making Penang cleaner, greener, and safer through education, technology, and initiatives. Be part of our family.

Stating that Penang is being made better via congruent measures is in keeping with the *logos* theme. The use of logic and reasoning here not only demonstrates the appropriateness of the council's decisions but also justifies the council's place/position within its immediate social system. The second statement falls under *pathos* and will be discussed in a later segment.

We present a series of other examples taken specifically from PGC's discourse on its green initiatives (https://pgc.com.my/2020/penang-green-initiatives/):

Extract L2

> The fact that 6.3 billion metric tonnes out of 8.3 billion metric tonnes of plastic produced end up in landfills or oceans has spurred our eco-conscious State Government to instate stringent (and widely successful) policies against it.

Extract L3

> Single-Use plastics are so harmful that the United Nations has launched a five years Clean Seas Campaign against it. It threatens not only aquatic life but also humans who drink from plastic bottles containing BPA (Bisphenol A), which can lead to cancers and increased miscarriage risk.

Extract L4

(Five individual threads are grouped together under this extract; they share the theme of why people need to segregate waste at source.)

> One Penangite generates 1.1 kg of waste daily. … In 1 day your waste is as heavy as 9 cats. … Penang had 1.767 million people in 2018. That's more than 1900 tonnes of waste in one day. … All waste in Penang go to Pulau Burung Landfill. … It costs the state RM5,000,000.00 to dump your waste in 2017 alone.

Extract L5

(Three individual threads are grouped together under this extract; they share the same theme: objectives of waste segregation at source.)

> To extend the lifespan of PB Landfill by cutting down waste generation. … To improve the cost-efficiency of waste management. … To lower pollution rates & maintain ecology balance.

On the whole, it is fair to surmise that PGC mainly utilised logic, reasoning, and factual statements to persuade. This practice also portrays the council as grounded in rationale and a fitting caretaker of the state's wellness in environmental terms. Also observed is an overlapping of appeals; for instance, Extract L2 is an appeal to logic but, at the same time, speaks of credibility within the context of the state government being eco-conscious and having the ability to act effectively.

Ethos

As mentioned, Extract L2 can be viewed as part of *ethos*. Extract E1 is another example:

Extract E1

> The State Government has implemented the No Single-Use Plastic initiative in Penang since 2018 in line with Malaysia's Roadmap towards Zero Single-Use Plastics 2018–2030. The Penang state government recently conducted a public survey on F&B operators' and the public's thoughts on no single-use plastic items in Penang's food eateries.

The first statement demonstrates the state government's responsible action in implementing the No Single-Use Plastic initiative and its willingness to fall in line with a national agenda, one that is timely and likely to benefit many. The second statement depicts the state government as a caring entity, receptive to the views of the people under its charge. As a whole, this approach can be appealing to the masses, and this can, in turn, persuade them to act in a certain way. Fundamentally, any stage set for persuasion is generally well served when *ethos* features noticeably in public discourse.

In addition, PGC's discourse taps into the realm of ingratiation (instrumental dependency), emphasising the importance of functioning together as a team. Extracts E2, E3 (concerning waste segregation at source), and E4 (PGC's mission statement) illustrate this.

Extract E2

> 60% of waste transported to Pulau Burung Landfill is recyclable through COMMITTED segregation.

Extract E3

> Your action makes a difference. Do your part today.

Extract E4

> To enable, empower, and enrich all stakeholders in practising sustainable development that protects the environment and quality of life.

Pathos

Aside from logic and ethical conduct, PGC's communication does showcase the emotional *pathos* as well. Within the context of its green initiatives discourse, manifestations of *pathos* are found in isolation and also in overlapping cases. With respect to the former, the second part of Extract L1 – "Be part of our family" – can be construed to embody some sense of *pathos*, as an emotive expression with familial substance may move the target audience to some degree, but perhaps not likely to the extent of inducing intense emotions.

Another expression that may be seen as embodying some sense of *pathos* is Extract P1, relating to waste segregation:

Extract P1

> There is hope with actions.

The expression calls for action from the audience – individuals, businesses, and institutions – and the call is paired with the word "hope", posing a statement with a certain margin of emotional value because the word is often associated with dreams, aspirations, optimism, and desirable outcomes. Additionally, in terms of overlapping cases, most of Extract L4 can also be viewed from the *pathos* perspective, in that most of the threads can potentially induce feelings of guilt that may then provoke eco-conscious actions.

Conclusion

We are each a social intervener through our rhetoric, and this becomes more obvious when communication is meant to persuade the masses. We consider the Aristotelian appeals to be the roots of research in persuasion and have included the notions of *logos*, *ethos*, and *pathos* in our analysis of selected texts from Penang Green Council's public discourse on sustainability. Our study is by no means exhaustive, as no study can be, but does help to extend the current knowledge base on sustainability communication in Malaysia.

All three appeals are manifested in the corpus that we elected to focus on. *Logos* features prominently in PGC's discourse, primarily in the use of facts, to steer the public toward green efforts. This is largely expected, and especially so in sustainability communication, because evidence relating to the environment (statistical data and actual events or incidents) is usually available in abundance. However, there is one matter worthy of note: as can be seen in Extract L4, there is use of data pertaining to 2017 and 2018 but none specified to pertain to more recent years. This is rather surprising given that the ability to provide more contemporary data is likely to portray PGC in a better light.

The employment of *ethos* is also evident in PGC's green initiatives discourse, with an inclination toward the application of instrumental dependency. It is possible that this inclination is due to the reasoning that the notion of having the power to make a difference can be attractive to the human psyche, and that citizens normally feel a sense of pride and allegiance to their hometown, state, and/or nation. Applying instrumental dependency also portrays PGC in an ethical light, demonstrating its humility, sense of practicality, and respect for the *rakyat* by including them in problem-solving efforts.

Pathos makes its appearance in the analysed corpus with emotive expressions that may move the target audience to some degree, but perhaps not likely to the point of evoking intense emotions. The disinclination to push harder on emotions may be due to the preference to not risk frustrating the audience. After all, in the face of problems, there is often a greater collective need for rationale and reliability instead of soul-stirring statements. In this case, a greenhushing technique that mitigates consumer guilt is applicable.

PGC's green initiatives seem to have struck a chord with existing literature, that green communication must be personalised, transparent, and clear and must motivate individuals' action (Kronrod et al., 2012; Stibbe, 2009). Rather than simply being politically correct, drastic alternative discourses are sometimes needed to instil immediate awareness and instigate action by citizens, but this will only work when the public is passionate about an issue. Essentially, any sustainability communication must observe societal and cultural values and norms to, first, attract the audience to engage with the green messages and not deter their interest. Then, more persuasive, emotional, and appealing messages may progressively transform their behaviour into a more sustainable one.

Acknowledgement

Shaidatul Akma Adi Kasuma is the corresponding author.

References

Abdul Rahman, H. (2020). Environmental sustainability awareness in selected countries. *International Journal of Academic Research in Business & Social Sciences, 10*(15), 85–97.

Abdullah, A. N., Husain, K., & Husin, N. (2009). Environmental communication and sustainable development in Malaysia: An analysis on environmental NGOs websites design. *Journal of Human Capital Development (JHCD), 2*(1), 37–52.

Allen, M. (2017). *The SAGE encyclopedia of communication research methods*. California: SAGE.

Anderson, A. A., Scheufele, D. A., Brossard, D., & Corley, E. A. (2012). The role of media and deference to scientific authority in cultivating trust in sources of information about emerging technology. *International Journal of Public Opinion Research, 24*(2), 225–237.

Bailey, A., Giangola, L., & Boykoff, M. T. (2014). How grammatical choice shapes media representations of climate (un)certainty. *Environmental Communication, 8*(2), 197–215.

Bakir, V., Herring, E., Miller, D., & Robinson, P. (2019). Organized persuasive communication: A new conceptual framework for research on public relations, propaganda and promotional culture. *Critical Sociology, 45*(3), 311–328.

Bergmann, Z., & Ossewaarde, R. (2020). Youth climate activists meet environmental governance: Ageist depictions of the FFF movement and Greta Thunberg in German newspaper coverage. *Journal of Multicultural Discourses, 15*(3), 267–290.

Brown, M. (2013). A methodology for mapping meanings in text-based sustainability communication. *Sustainability, 5*(6), 2457–2479.

Byrum, K. (2019). "Hey Friend, Buy Green": Social media use to influence eco-purchasing involvement. *Environmental Communication, 13*(2), 209–221.

Chestney, N., & Januta, A. (2021, August 10). *U.N. climate change report sounds 'code red for humanity'*. Reuters. Accessed from https://www.reuters.com/business/environment/un-sounds-clarion-call-over-irreversible-climate-impacts-by-humans-2021-08-09/

Cooper, D. E. (2001). Aristotle 384–322 BCE. In J. A. Palmer, D. E. Cooper & P. B. Corcoran (Eds.), *Fifty key thinkers on the environment* (pp. 12–17). London: Routledge.

Dovers, S. R., & Handmer, J. W. (1993). Contradictions in sustainability. *Environmental conservation, 20*(3), 217–222.

Genç, R. (2017). The importance of communication in sustainability & sustainable strategies. *Procedia Manufacturing, 8*, 511–516.

Giebelhausen, M., Chun, H. H., Cronin, J. J., & Hult, G. T. M. (2016). Adjusting the warm-glow thermostat: How incentivizing participation in voluntary green programs moderates their impact on service satisfaction. *Journal of Marketing, 80*(4), 56–71.

Godemann, J., & Michelsen, G. (2011). Sustainability communication: An introduction. In J. Godemann & G. Michelsen (Eds.), *Sustainability communication* (pp. 3–11). Dordrecht: Springer.

Grabowski, I. (2007). Consumed by consumerism: The persuasive discourse of financial institutions. *Language & Ecology, 2*(2), 1–15.

Green, M. C. (2006). Narratives and Cancer Communication. *Journal of Communication, 56*(1), S163–S183.

Hussain, W. (2017). *Transparency in sustainability communication: Developing and testing a new model to assess consumer-facing sustainability transparency communication of fashion brand/retailer's website* [Unpublished Master Thesis]. The Swedish School of Textiles: University of Boras.

Jha, A. K., & Verma, N. K. (2022). Social media sustainability communication: An analysis of firm behaviour and stakeholder responses. *Information Systems Frontiers*, March, 1–20.

Joosse, S., & Brydges, T. (2018). Blogging for sustainability: The intermediary role of personal green blogs in promoting sustainability. *Environmental Communication, 12*(5), 686–700.

Kapoor, P. S., Balaji, M. S., & Jiang, Y. (2021). Effectiveness of sustainability communication on social media: Role of message appeal and message source. *International Journal of Contemporary Hospitality Management, 33*(3), 949–972.

Koszowy, M., Budzynska, K., Pereira-Fariña, M., & Duthie, R. (2022). From theory of rhetoric to the practice of language use: The case of appeals to *ethos* elements. *Argumentation, 36*, 123–149.

Kronrod, A., Grinstein, A., & Wathieu, L. (2012). Go green! Should environmental messages be so assertive?. *Journal of Marketing, 76*(1), 95–102.

Lowrey, W., Evans, W, Gower, K. K., Robinson, J. A., Ginter, P. M., McCormic, L. C., & Abdolrasulnia, M. (2007). Effective media communication of disasters: Pressing problems and recommendations. *BMC Public Health*, 7, 97.

Luu, T. T. (2021). Can food waste behavior be managed within the B2B workplace and beyond? The roles of quality of green communication and dual mediation paths. *Industrial Marketing Management, 93*, 628–640.

Martin, C., & MacDonald, B. H. (2020). Using interpersonal communication strategies to encourage science conversation on social media. *PLOS ONE, 15*(11), e0241972.

Milanesi, M., Kyrdoda, Y., & Runfola, A. (2021). How do you depict sustainability? An analysis of images posted on Instagram by sustainable fashion companies. *Journal of Global Fashion Marketing, 13*(2), 101–115.

Mohamad Saleh, M. S. (2017). Media and environmental non-governmental organizations (ENGOs) roles in environmental sustainability communication in Malaysia. *Discourse and Communication for Sustainable Education, 8*(1), 90–101.

Mohamad Saleh, M. S., Adi Kasuma, S. A., Balaraman, R. A., & Syed Mukhiar, S. N. (2018). The current progress of Malaysian media and environmental non-governmental organisations (ENGOs) in environmental education: An evaluation. *Jurnal Komunikasi Borneo, 6*(Special Issue/December: 20th UMS Convocation), 38–49.

Mohamad Saleh, M. S., & Heinrichs, H. (2020). Media representation of environmental issues in Malaysia: Newpapers and environment non-government organisation newsletters. *Pacific Journalism Review, 26*(2), 118–131.

Mohamad Saleh, M. S., Heinrichs, H., & Nik Hasan, N. N. (2018). The struggles of Malaysian media and environmental non-governmental organisations (ENGOs) in communicating the environment within semi-democratic nation. *The Journal of International Communication, 24*(1), 55–75.

More, H. (2019). Stockholm declaration. *The fact factor.* Accessed from https://thefactfactor.com/facts/law/civil_law/environmental_laws/stockholm-declaration/871/

Nwagbara, U., & Belal, A. (2019). Persuasive language of responsible organisation? A critical discourse analysis of corporate social responsibility (CSR) reports of Nigerian oil companies. *Accounting, Auditing & Accountability Journal, 32*(8), 2395–2420.

Pekkala, K., & van Zoonen, W. (2022). Work-related social media use: The mediating role of social media communication self-efficacy. *European Management Journal, 40*(1), 67–76.

Penang Green Council. (2020). Accessed from Penang Green Council. Accessed from https://pgc.com.my/2020/

Reilly, A. H., & Hynan, K. A. (2014). Corporate communication, sustainability, and social media: It's not easy (really) being green. *Business Horizons, 57*(6), 747–758.

Rossi, J. A. (1996). Controversial Public Policy Issues-Other Lessons. *Social Education, 60*(1), 15–21.

Saran, S. M., & Shokouhyar, S. (2021). Crossing the chasm between green corporate image and green corporate identity: A text mining, social media-based case study on automakers. *Journal of Strategic Marketing*, DOI: 10.1080/0965254X.2021.1874490.

Smith, R. D. (2021). *Strategic planning for public relations: 6th Edition.* New York and London: Routledge Taylor & Francis Group.

Stibbe, A. (2009). *The language of sustainability: Shouting but not being heard* [Keynote Address]. London: Institution for Environmental Sciences.

Supran, G., & Oreskes, N. (2021). Rhetoric and frame analysis of ExxonMobil's climate change communications. *One Earth*, *4*(5), 696–719.

Tan, D. A. L., & Al-Rikaby, A. B. M. (2022, In Press). Aristotelian framing in political discourse: A case study of Malaysian Prime Minister Muhyiddin Yassin's exclusive interview with *Radio Televisyen Malaysia* and the news agency *Bernama*. In O. Feldman (Ed.), *Debating politics: Political interviews and the art of jousting in staged media performances* (pp. tba). Singapore: Springer Nature.

Tiago, F., Gil, A., Stemberger, S., & Borges-Tiago, T. (2021). Digital sustainability communication in tourism. *Journal of Innovation and Knowledge*, *6*(1), 27–34.

Tölkes, C. (2020). The role of sustainability communication in the attitude–behaviour gap of sustainable tourism. *Tourism and Hospitality Research*, *20*(1), 117–128.

Voci, D. (2022). Logos, ethos, pathos, sustainabilitos? About the role of media companies in reaching sustainable development. *Sustainability*, *14*, 1–17.

Westra, L., & Robinson, T. M. (1997). *The Greeks and the environment*. Maryland: Rowman & Littlefield.

Yacoumis, P. (2018). Making progress? Reproducing hegemony through discourses of "sustainable development" in the Australian News Media. *Environmental Communication*, *12*(6), 840–853.

Yeniçeri Alemdar, M., & Kocaömer, C. (2020). Reading instagram posts of environmental nongovernmental organizations in the context of dialogic principles and public relations models. *Connectist: Istanbul University Journal of Communication Sciences*, *59*, 299–331.

6

IDEATING SUSTAINABILITY CONTENT IN SOCIAL MEDIA

Nur Atikah A Rahman

Introduction

Recently, there has been a rising demand in the market for sustainable consumption of products and services that are environmentally friendly. Based on past research, the COVID-19 pandemic played a part in this increased demand for sustainability consumption. The term sustainability consumption refers to when consumers and corporations focus on addressing ecological issues by purchasing and utilising products and services to preserve the environment (Ying & Shanyong, 2020) with societal impacts. Shrum et al. (1995) said that consumers who purchase sustainable products are also recognised as green consumers, those that have the awareness to change their wasteful or damaging consumption patterns. These green consumers prefer eco-friendly products and services with ecological sensibility. According to Parker (2021), 80% of global citizens were willing to choose sustainable brands, products, or services. Sustainable products or services are defined as goods and services that have harmless ingredients, are organically grown and biodegradable, have not been tested on living animals, and so on (Ottman, 1993; Pavan, 2010; Norazah, 2013).

To be explicit, according to global research conducted in 2021 by the Economist Intelligence Unit (EIU) and commissioned by the World Wild Fund for Nature (WWF), there was a 71% increase in consumers looking for sustainable products over five years. Most online searches for sustainable products can be seen in countries such as the United States, United Kingdom, Germany, Australia, and Canada, followed by other countries such as Ecuador and Indonesia. This contrasts with data from before the pandemic, where Chinese and Indian consumers were proactive in green consumption, whereas Australian, Japanese, and Korean consumers were the least proactive (Lung, 2010). The WWF (2021) report also stated that the COVID-19 pandemic was one of the reasons for the emerging

DOI: 10.4324/9781003309642-6

market and interest in sustainable consumption. Owing to the health risks that consumers are now facing because of COVID-19, many consumers make consumption decisions based on environmental factors to be healthier and live longer. In the United States, the Adweek Sustainability Report (2022) indicated that, in 2021, Twitter discussions concerning sustainable growth and decarbonisation rose to 53% and 56%, respectively. Similarly, Parker (2021) connoted that there is an increasing awareness of sustainability practices among Asian consumers. During the pandemic, about 58% of these consumers were personally impacted by environmental problems. Consequently, 53% consumers have stopped purchasing from brands that affect the environment and society negatively.

In the same year, GreenPrint's (2021) Business of Sustainability Index Report indicated that 75% of millennials in the United States are willing to fork out more money to purchase sustainable products, followed by Gen Z (63%), Gen X (64%), and boomers (57%). Likewise, about 80% of Thai, Malaysian, and Korean consumers are also willing to spend a premium to buy sustainable products in the market (Dunlap & Scarce, 1991; Lung 2010; Norazah, 2013). But, fewer than 60% of consumers in Hong Kong and Australia were likely to spend more on green products or services (Lung, 2010). Moreover, based on the findings of Norazah (2013), Generation Y in Malaysia was a young community that also showed responsiveness towards a sustainable environment. These statistics showed that consumers are now progressing towards making a greener effort in their purchase intentions as well as buying decisions. Thus, corporations are now competitively following this trend to uphold sustainable efforts, as they see that this is a lucrative opportunity to increase the return on investment (ROI) for their brands and businesses. In brief, when corporations fulfil consumers' demands, they will be easily influenced in terms of purchase intention and decisions.

With the increasing demand for sustainable products and services, the adoption of sustainability advertising is showing a growing trend in the discipline of brand communication. Advertising plays a huge role in communicating eco-friendly messages to foster sustainable awareness among consumers (Grillo et al., 2008; Leonidou et al., 2011, p. 6). This approach promotes environmentally and socially responsible brands, products, or services in the market (Story, 2019). Segev et al. (2016, p. 106) noted that consumers are now exposed to a mounting volume of sustainability messages and content, igniting "climate guilt" about how their daily life or lifestyle has affected the environment and society. Thus, corporations are having a hard time standing out among the saturated noise where consumers perceive those advertisements as disruption of information. Today, consumers deny interruptive content, making it harder for them to trust and understand messages that promote these initiatives.

Based on past research, there were diverse kinds of responses towards sustainability advertising, whether positive or negative. Positively, consumers are now aware of the products and services offered and are moving towards sustainability consumption. However, there is also disbelief (Carlson et al., 1993) and scepticism (Kim et al., 2017) towards sustainability advertising and corporations. This was also supported by Ongkrutraksa (2002), who stated that, based on Dentsu's report in 1993, while sustainability advertisements have garnered high consumer recognition, consumers are still lacking trust in the message, as the amount of sustainable content was considered insufficient. Some consumers had positive perceptions, indicating that sustainability advertising was portraying their responsibility towards the environment and showed sincerity, while the negative perception was more how they perceived that these corporations who advertised their green efforts were considered deceitful and shallow (Ongkrutraksa, 2002).

According to Matthes et al. (2014), one of the drawbacks of inconsistent and disapproving content is when it is perceived as greenwashing. Deskera (2022) opined that consumers want to trust that what they are purchasing is actually eco-friendly and sustainable. Greenwashing is when corporations promote products that are environmentally friendly to consumers but, in reality, they have little or no positive impact on the sustainable movement (Deskera, 2022). This is also supported by GreenPrint (2021), which mentioned that most Americans are suspicious about the corporate sustainability messages that organisations display. Specifically, in 2021, there was an increase of 158% of tweets mentioning greenwashing advertisements (Adweek Sustainability Report, 2022).

Therefore, corporations have to be true to their sustainable initiatives and portray the right kind of content and messages. Drawing on this purpose, corporations are now looking into branded and strategic approaches to reach consumers and engage them to act upon the messages portrayed – which is to purchase their sustainable product and service or engage in any societal cause. The brand should look, sound, and feel truly green. So, how do corporations ensure that they can foster trust among consumers in order for them to change their minds and behaviour?

To answer this question, the content of digital experiences such as social media content plays an important role in facilitating a relationship with consumers as well as in forming favourable attitudes and behaviour towards certain green brands, products, or services (Ying & Shanyong, 2020) as well as humans' responsibility for economic and societal impacts. However, this can only be effectively executed if corporations employ the right strategy and content to capture their consumers' attention and influence their purchasing intention. Recently, social media have become important channels for organisations to communicate sustainability practices (Pittman et al., 2022). It is perceived that sustainability advertising on social media is considered trustworthy and credible, where 80% of the consumers are aware of green marketing on social media (Abzari et al., 2014; Ying & Shanyong, 2020).

In this case, social media and sustainability go hand in hand in brand communication strategies, helping corporations raise awareness of their efforts, generate interactivity, and foster greater engagement with consumers. However, Ying and Shanyong (2020) also indicated that, although social media play a pivotal role in sustainability marketing, not all consumers will purchase the products or services or engage in the sustainability practices immediately after receiving the information. Therefore, corporations or content strategists need to look further into the structure, strategy, and storyline of the content to ensure that it is effective and resonates with consumers.

Sustainability Advertising

Sustainability communication is the act of delivering messages to consumers about a certain product or service, promoting its unique features and benefits to meet consumers' demands for a better environment and life (Kapoor et al., 2021). According to the Adweek Sustainability Report (2022), sustainability marketing is referred to as "business practices that take into account the long-term impacts on the environment and human society rooted in a strong understanding of climate science" (p. 4). Kirchgeorg and Winn (2006) defined sustainability marketing as "the planning, coordination, implementation and controlling of all market transactions" that consider ecological and social impacts (p. 176).

Sustainability advertising is known as messages or content that are used to inform and promote environmental aspects of brands, products, or services, to change consumers' minds and behaviours in favour of eco-friendly consumption and sustainable issues (Han et al., 2019). According to Minton et al. (2012) as cited in Han et al. (2019), sustainability advertising is used interchangeably with terms such as green advertising. Based on the definition of Banerjee et al. (1995), green advertising is when a corporation produces brand communication materials that address ecological issues and relate their brands, products, or services as a solution to sustain the environment. Green advertising was first introduced in the 1970s, owing to the recession that was caused by high fuel prices and increasing environmental issues (Haytko & Matulich, 2008). Zinkhan and Carlson (1995) inferred that green advertisements are defined as promotional messages that focus on environmental sustainability, consisting of three major categories such as (Leonidou et al., 2011):

1. Direct or indirect discussion of how the products or services contribute to the natural environment;
2. Promotion of a green lifestyle by using or avoiding using the product or service;
3. Portrayal of a pro-environmental image as part of the corporation's corporate responsibility.

Across Asia, there have been several green marketing approaches to ensure that people can drive their purchasing choices towards a greener selection. Particularly in Japan, green advertising has gained public interest since the 1990s, with Volvo's print advertisement in *Nikkei Newspaper (Nihon Keizai*

Shimbun, 1994; Ongkrutraksa, 2002). In the same way, the popularity of green advertising in Thailand has also risen since the 1990s, where it focuses on Thai's air pollution phenomenon (Saereerat, 1995; Ongkrutraksa, 2002). In promoting more on sustainability communication, multinational companies (i.e. Caltex, Shell, Mercedes Benz, Honda, Nissan) and the Petroleum Authority of Thailand employ a series of green advertising campaigns (Saereerat, 1995; Ongkrutraksa, 2002).

In contrast, in China, sustainable consumption and consumerism started later than other countries and came into effect owing to a series of environmental issues caused by the Open Door Policy and economic transformation (Zhu & Sarkis, 2016). According to Yu (2014), consumers in China started to change their environmental attitudes and consumption of sustainable products to overcome declines in environmental performance such as emissions of SO_2 and CO_2 as well as freshwater pollution.

As for other Asian countries, South Korea has changed its policies to expand the green movement by adopting energy-saving solutions and eco-friendly consumption (Dae Ryun, 2012; Sathiswaran et al., 2014). In Malaysia, the nation's commitment to a "Green Malaysia" was fully launched in 2009 with the emergence of the Malaysian Green Technology Policy. Since then, countless environmental campaigns have been conducted, including No Plastic Bag Day, Sime Darby's Plant a Tree Program, Digi's Mangrove-Saving Project, and more (Mohd Helmi et al., 2012).

Greenwashing in Sustainability Advertisements

Although sustainability advertising is increasingly used by corporations and brands, consumers are also showing signs of scepticism about content that promotes a pro-environmental image and consciousness in the market (Leonidou et al., 2011). Consumers from around the world have specified that sustainability advertising lacks credibility and trustworthiness, causing scepticism. Paco and Reis (2012) explained that scepticism can undermine the positive influence of sustainability advertising on consumers. Knight (2008) stated that consumers are concerned with the rising number of advertisements that are green-themed and have misleading claims in their content. This has led to the identification of sustainability advertisements as greenwashing.

According to Deskera (2022), greenwashing is when corporations use advertisements to promote their environmental message, but the advertising claims are considered overstated and have little or no impact on green activity. In specific, greenwashing happens when the advertisement has ambiguous phrases in the content, excludes facts or accurate data, and contains misleading information (Kangun et al., 1991; Leonidou et al., 2011).

For example, in 2008, the Malaysian Palm Oil Council created a television commercial that promoted its environmental initiatives. The content of the commercial was on "Malaysia Palm Oil. Its trees give life and help our planet breathe, give homes to hundreds of species of flora and fauna. Malaysia Palm Oil. A gift from nature, a gift for life" (Dahl, 2010, p. 248). Unfortunately, Friends of the Earth Europe made a complaint about the advertisement as it believes that palm oil plantations lead to animal extinction, loss of habitat, pollution, floods, and other ecological problems. This is the absolute opposite of a sustainable environment for the country. This resulted in the UK Advertising Standards Authority (ASA), a UK advertising watchdog, to rule that the commercial offered false advertising or greenwashing as palm oil is not sustainably produced. Moreover, it was judged to use deceptive arguments and images with the slogan of "a gift from nature, a gift from life" and videos of the lush foliage of the rainforest (Friends of the Earth Europe, 2012). However, the Malaysian Palm Oil Council disagreed with the ruling of the ASA as the Malaysian Palm Oil Industry does have a key role in decreasing poverty in the country, and the oil was sustainably produced with verification of its sustainability from an organisation sponsored by WWF (MPOC, 2009).

In another example, Fernando, Sivakumaran, and Suganthi (2014) stated that sustainability advertisements should avoid using vague claims that contain confusing promises about a corporation or brand. For instance, in India, an organisation used the term, "We are a green company". This is an example of a shallow claim that offers no direct meaning of what it is trying to say pertaining to its green cause. To make this content carry a better environmental message, content developers could change it into concrete claims such as "Over the last few years, we have reduced the weight of the bottle by 22%. And today, it weighs 18.5 grams, shrinking the amount of plastic used" (Fernando et al., 2014).

Apart from that, Tolkes (2020) has indicated that sustainability communication is sometimes ineffective because consumers are not able to process the messages. Consumers then become very suspicious of every message and all content that is related to sustainability communication (Carlson et al., 1993). Some of the words used in green advertising content that can be considered vague are terms such as "100% natural", "earth-friendly", "energy efficiency", "energy conservation", and more, which do not really portray what they mean in relation to sustainable initiatives (Knight, 2008; Murphy et al., 2010). Paco and Reis (2012) also stated that some of the unclear messages use language such as "biodegradable", "environmentally friendly", and "ozone friendly". Use of these words has led to green advertisements being perceived as too complicated, "jargon-ish", or misleading for consumers to understand their message.

Thus, preventing sustainable advertisements from portraying greenwashing messages can further assist in developing trust among green consumers (Chen & Chang, 2012). In order to do so, corporations and content strategists must look at their content to ensure that the environmental claims conveyed in those messages are valid, credible, and trustworthy in order to enlist their targeted consumers in supporting pro-environment causes.

Sustainability Social Media Advertising

In the digitised world of today, communication technology such as social media has tremendously changed our consumption of information and interaction with others. Specifically, social media have been adopted as an effective cyberspace in which to disseminate information, foster relationships with other users online, and interact and share content with consumers (Ying & Shanyong, 2020). According to Reilly and Hynan (2014, p. 749), social media are known as "technology-facilitated dialogue" that is distributed via social networking sites, blogs and microblogging, content communities, social bookmarking, and more. Moreover, We Are Social (2022) reported that, in January 2022, 4.62 billion people (58.4% of the total population) were active on social media around the world. The report also described Facebook as the world's most-used social platform, followed by YouTube, Instagram, TikTok, Douyin, Snapchat, Pinterest, Twitter, and more (We Are Social, 2022).

In the business context, social media have advanced rapidly, allowing corporations to adopt innovative and engaging platforms as part of their brand communication to attract and persuade consumers to purchase or acquire their brand, products, or services. Among some of the popular social media platforms that offer advertising and brand positioning features are social networking sites (such as Facebook, Instagram, Tik Tok, LinkedIn), blogs and microblogging platforms (such as WordPress, BlogSpot, or Twitter), content communities (such as YouTube), and more. These platforms enable consumers to interact with other consumers who have the same environmental interests and goals (Ying & Shanyong, 2020).

For example, on Instagram alone, the total potential reach of advertisements on Instagram is 1.48 billion users, where the home feed reaches about 96.6% of its active users, advertisements in Instagram Stories reach more than 1 billion users, and reels reach 675.3 million users each month (We Are Social, 2022). Data from the report also suggested that TikTok advertising is on the rise across the world, with a potential reach of 884.9 million users aged 18 years and above. What is more interesting is that the reach of TikTok is taking over Asian countries such as Thailand, Malaysia, Cambodia, Vietnam, the Philippines, Indonesia, and more (We Are Social, 2022).

According to Pittman et al. (2022), to ensure that corporations are able to capture the interest and minds of consumers, careful tactics and context must be considered to ideate the right content for their brand, product, or services. On social media, consumers are also more aware of their identity or self-presentation to be accepted and liked by sustainable consumption groups. Thus, what they hear, say, and do is influenced by the people surrounding them. Ying and Shanyong (2020) also stated that social media users have the choice to obtain certain information as well as post and share their own content pertaining to preserving nature and the environment online. They also make their own purchase decisions based on the information on social media.

In addition, Ying and Shanyong (2020) also revealed that social media have personalisation features that make them useful and reliable tools for promoting sustainability content as well as fostering consumers' intention to adopt pro-environmental behaviour. Moreover, social media marketing also helps consumers gain more product knowledge, affecting their product awareness and price consciousness. Froehlich (2009) indicated that social media marketing also aids in convincing consumers about sustainable information to encourage preservation of the environment.

Content Ideation for Sustainability

According to Van den Ende et al. (2014), ideation is one of the most vital processes in creating innovation, whereas content is the text, pictures, audio, or video that corporations disseminate to their target consumers. In particular, content ideation is the strategic process of crafting compelling and effective messages through text, imagery, audio, and video to achieve the marketing and communication objectives of the corporation. In ideating content, the message should be able to move specific audiences towards acquiring and purchasing sustainable products or services. Positioning and promoting a corporation or brand as a "green brand" require strategic brand communication to ensure that it offers benefits to the environment. According to Pulizzi (2014), the content created should be engaging, informative, and attractive to accomplish the business goals.

In particular, sustainable content in social media advertising should not only focus on selling products and services fast, but should also generate curiosity and build trust and association between the corporation and consumers (du Plessis, 2015). Based on past literature, sustainability content is different from other content, which may affect its persuasiveness (Randle et al., 2019). Sustainability content utilises different language, addresses several distinct environmental issues, and has a diverse kind of structure that may be considered complicated to understand. Thus, to increase the reach of the content, corporations and content strategists should nurture consumers to buy the products and services through the right message appeal.

Message Appeals in Sustainability Content

In sustainability advertisements, the content used a wide array of message appeals that offered various tones of expression, such as rational and emotional. Message appeal is the overall theme of sustainability content that can help to persuade consumers to go for an eco-friendly choice (Wang & Lehto, 2019).

Rational appeal is when the content is crafted to give practical and utilitarian information about the product's attributes (Hartmann et al., 2005). For example, in rational appeal sustainability advertisements, the content usually delivers messages that are related to the facts of how air pollution can be reduced if emissions are reduced. There are no emotional connotations (Matthes et al., 2014). Based on the analysis by Leonidou et al. (2011) of international sustainability advertisements, about 99.7% of sustainability advertising utilised a rational approach in the body copy. However, Matthes et al. (2014) further stated that rational appeal content does not affect all types of consumers and is more effective with consumers who are frequent buyers of green products or services.

On the other hand, about 34.3% of sustainability advertising utilised emotional appeals (Leonidou et al., 2011). In emotional appeals, the content is created to offer messages that connect with the target audiences' attitude and feelings. An emotional strategy helps to make content more personal and relevant, driving attraction and trustworthiness (Marsden & Chaney, 2013). To drive emotions and empathy, corporations need to involve their audience with the content or product.

According to Matthes et al. (2014), emotional appeal content is vital in shaping consumers' attitudes to form a close relationship with real nature (Hartmann & Apaolaza-Ibanez, 2009). Similarly, in Malaysia, Chen and Chai (2010) stated that advertisements that develop and implement emotional messages are effective in boosting greater sales of sustainable products. Moreover, Kapoor et al. (2021) indicated that emotional appeals are considered more impactful than rational appeals, especially for hedonic types of services.

To be precise, an emotional appeal accentuates hedonic experiences such as guilt, fear, adventure, romance, and sex (Kapoor et al., 2021, p. 953). Likewise, Leonidou et al. (2011) found the emotional appeals used most frequently are those that target consumers' pleasure or pride. More importantly, an emotional appeal consists of two valences, positive and negative (Kapoor et al., 2021). A positive emotional appeal includes sensual, humorous, and social elements that have a greater impact on consumers' behavioural intention to purchase hedonic products (Ekici et al., 2020). In fact, sensual appeal focuses on sensuality, pleasure, and visual cues to gratify the senses (Brito & Pratas, 2015). Sensual appeals are known for their messages that are engaging, stimulating, and entertaining (Ekici et al., 2020; Mayer et al., 2019; Shen et al., 2020). For example, in Kapoor et al.'s (2021) study of sustainability advertising on social media, the use of sensual message appeals for hotels is more effective to drive intentions to stay overnight at an environmentally friendly hotel. Sensual messages include cues such as a romantic atmosphere, stunning views of the rainforest, and more.

In contrast, negative emotional appeals consist of guilt, shame, and fear (Kapoor et al., 2021) in sustainability communication that invokes self-protection. According to Coleman et al. (2020), a guilt appeal can also help to stimulate consumers to form pro-sustainable intentions and decisions. Guilt is usually classified as emotions such as repentance and self-blame in relation to sustainable conditions that will motivate consumers to react to it. Based on content analysis of green advertising in Japan, Thailand, China, and the United States, there are several kinds of message appeals that have been used in portraying sustainable communication and advertising for consumers (Ongkrutraksa, 2002). In Japan, most of the sustainability advertising employs a softer-sell approach using positive emotional appeals such as reward appeal. This is because the Japanese are considered a high-context society, where logical reasoning may not work, compared with a consumer-oriented approach. Ongkrutraksa (2002) also mentioned that, unlike Japan, Thailand's sustainability advertising used negative emotional appeals such as a fear appeal. Hence, it is crucial for content to be crafted with a consumer-centric approach to add value to the lives of consumers. Based on these findings above, it can be inferred that sustainability content in Asia adopts emotional appeals to resonate with consumers. However, whether a corporation or brand uses a positive or negative emotional appeal is entirely based on the consumers' persona and empathy map, to ensure that the content is relevant and suitable for their sustainable wants and needs.

Elements of Effective Content Ideation

To engage wider audiences, corporations need to know how to create sustainability-related dialogues. In reality, people love stories, and one of the most effective content strategies is utilising the storytelling technique. Storytelling is utilised in brand communication to form relationships with consumers and convey complex green messages to those consumers (du Plessis, 2015). To capture the attention of the target audience (consumers), the storyline of the content should fascinate them and invoke emotions and feelings that foster ecological sensibility.

Strategic sustainable content should be able to move the audience to perceive that the message portrayed is targeted at them, personalised for their needs and problems, as well as worth their consumption. It is considered worth consuming when consumers actually read, listen, watch, and share the sustainability content. The storytelling of the content is conceptualised with a combination of the marketing and communication objective, product or service benefit, and creative ideas that are tailored to different types of consumers (du Plessis, 2015).

According to Bender (2020), narrating a sustainable storyline should start with using the customer as the hero, and the green product or service as the solution to overcome the ecological issues. In relation to that, it is also best for content strategists to understand their audiences' empathy map and buyer personas to develop content that focuses on their pains and gains when offering the right solution in the content. According to Engeseth (2006, p. 62), consumers are the main source of and contribution to a brand's content. Similarly, Pulizzi (2014) also indicated that content strategists should map the audience personas with their buying cycle. The buying cycle consists of processes such as green awareness, information search, request for pricing/product information, purchase decision, competitive alternative, and short-listing of solutions for their needs (Pulizzi, 2014).

In ideating sustainable content, it is suggested that content writers use the AIDA formula, which was developed in 1898 by Elias St. Elmo Lewis for his storytelling. Specifically, the AIDA model contains four steps to persuade consumers, which are to attract attention, spark interest, create desire, and take action. Content of sustainability advertising placed on social media must be able to attract the attention of the audience, capture their interest in knowing more about the product or service mentioned, create desire among the audience to make purchase intentions, and move towards taking action to purchase those sustainable products.

First, content strategists should start with a compelling headline that has a powerful and creative hook to instantly grab consumers' attention and direct it towards the sustainable product or service (Shaw, 2009). The first line should command their attention towards a point or anchor of the story. In order to do that, content strategists need to discover audiences' sustainable passion, aspirations, and problems regarding the environment. The headline should have the ability to distract audiences from the saturated sustainability information found in the market.

According to Leonidou et al. (2011), headlines with a declarative form and statements that have clearly defined meanings were employed most by international firms (75.5%), followed by identification (18.6%) and interrogative (17.1%) headlines. Moreover, the study also noted that some advertisements employed statistics in headlines to sensationalise the ecological issue. This is then followed by headlines such as quotations – selective and negative. To ideate headlines on social media, content strategists can also consider generating headlines that can be found through search engine keyword searches (search engine optimisation) with lead characteristics that can convert into sales (Pulizzi, 2014). The content must also be able to hold attention longer and reach as many customers as possible.

Then, content strategists should work on igniting the interest of consumers by holding their attention to incite curiosity and engagement. Interest can be created through the subheading of the storyline to pull the audience into the content and promise them exciting discoveries in the following. Sustainability content should be hooked with an engaging story that is consumer-oriented, and so strategists should be human and avoid any jargon that could tend towards sales or marketing (Pulizzi, 2014).

To create desire among consumers, sustainability body copy should have a single focused point. This section should deliver the promise by relating the revelations to the promise in the subheading (Shaw, 2009). The goal that the content is trying to inform and reach is to ensure that the consumer moves through the customer journey to purchase sustainable product or services. Well-crafted stories can pull consumers from experiencing pain concerning ecological issues to feeling the gain from the benefits of the product or services offered.

According to Leonidou et al. (2011), 50.1% of body copy used descriptions of key issues, followed by an explanative approach (47.8%). Furthermore, Leonidou et al.'s (2011) analysis also found that factual, persuasive body copy is perceived to offer strong, relevant, and focused statements about ecological issues. Therefore,

Leonidou et al. (2011) suggested that, to create effective and relevant messages for sustainability storytelling, a content strategist should focus on:

1. Offering comprehensive information on the ecological concern;
2. Focusing on the environmental benefits that consumers can reap from the product's or service's adoption of green initiatives;
3. Highlighting the corporation's green activities and events to clearly distinguish themselves from the competitors that do not offer any green benefits.

Additionally, compelling facts and, sometimes, humour can be embedded in body copy. At times, humour can help to make content less serious and can resonate with the audience, fostering their interest. Nonetheless, at the same time, content strategists should not overdo the humour, as it can have subjective meanings for consumers (Shaw, 2009). Moreover, too much humour can also make the sustainability content sound too relaxed and untrustworthy.

Lastly, the content strategist crafts call-to-action statements to lead consumers towards a certain action for future purchases. In social media, call-to-action statements can invite consumers to visit a website (Shaw, 2009), play an augmented reality game, or click on the advertisement for conversion into a sale. Corporations and content strategists can also create engagement with audiences by asking them to act by saving and sharing the content. It is also advisable for content strategists to integrate specific information such as compelling hashtags or augmented reality games to get the consumers to respond and interact with the content. Past studies have also indicated that social media should be used not only as an advertising channel but also to get consumers' engagement and interaction with the corporation or sustainable brand (Pulizzi, 2014). As part of brand engagement, there are several categories of activities that lead consumers to engage with ads on social media, by instilling content that encourages consumers to like, share, save, and comment on the content.

Conclusion

In summary, sustainability advertising is a strategic approach for corporations to disseminate information and promote ecological and sustainable causes. However, as the communication paradigm changes owing to the emergence of social media marketing, consumers are no longer passive receivers of messages and are proactively reacting to messages, whether positively or negatively. Thus, corporations and content strategists need to work diligently and effectively to differentiate their content from other competitors in the market, while ensuring that their green advertising is not labelled as greenwashing. This chapter has discussed, in particular, strategic approaches to content marketing on social media by explaining message appeals, sustainability storytelling, and elements of sustainability content.

References

Abzari, M., Ghassemi, R. A., & Vosta, L. N. (2014). Analyzing the effect of social media on brand attitude and purchase intention: The case of Iran Khodro Company. *Procedia-Social and Behavioural Sciences, 143*(14), 822–826.

Adweek Sustainability Report. (2022). The climate is changing. Are you? How sustainability is redefining the relationship among agencies, brands, consumers and the media. *Adweek*. Accessed from https://www.adweek.com/the-climate-is-changing-are-you/

Banerjee, S., Gulas, S., & Iyer, E. (1995). Shades of green: A multidimensional analysis of environmental advertising. *Journal of Advertising, 23*(Summer), 21–31.

Bender, A. (2020). *Creative content kit: Ideate and create content strategy*. Amsterdam: BIS Publishers.

Brito, P. Q., & Pratas, J. (2015). Tourism brochures linking message strategies, tactics and brand destination attributes. *Tourism Management, 48*, 123–138.

Carlson, L., Grove, S. J., & Kangun, N. (1993). A content analysis of environmental advertising claims: A matrix method approach. *Journal of Advertising, 22*(3), 27–39.

Chen, T. B., & Chai, L. T. (2010). Attitude towards the environment and green products: Consumers perspective. *Management and Science Engineering, 4*, 27–39.

Chen, Y. S., & Chang, C. H. (2012). Greenwash and green trust: The mediation effects of green consumer confusion and green perceived risk. *Journal of Business Ethics, 114*, 489–500.

Coleman, J. T., Royne, M. B., & Punders, K. R. (2020). Pride, guilt and self-regulation in cause-related marketing advertisements. *Journal of Advertising, 49*(1), 34–60.

Dae Ryun, C. E. K., (2012). Increasing the impact of sponsorship on brands and reputation. *International Journal of Advertising, 31*(1), 12–14.

Dahl, R. (2010). Green washing. *Environmental Health Perspectives, 118*(6), 246–252.

Deskera. (2022). The complete guide to green marketing. Accessed from https://www.deskera.com/blog/green-marketing/

du Plessis, C. (2015). Brand storytelling: The case of Coca-cola's journey corporate website. *Communitas, 20*, 84–103.

Dunlap, R., & Scarce, R. (1991). Environmental problems and protections. *Public Opinion Quarterly, 55*, 651–672.

Ekici, N., Erdogan, B. Z., & Basil, M. (2020). The third-person perception of sex appeals in hedonic and utilitarian product ads. *Journal of International Consumer Marketing, 32*(4), 336–351.

Engeseth. (2006). *ONE: A consumer revolution for business*. New Industrial Road, Singapore: Marshall Cavendish Business.

Fernando, A. G., Sivakumaran, B., & Suganthi, L. (2014). Nature of green advertisements in India: Are they greenwashed? *Asian Journal of Communication, 24*(3), 222–241.

Friends of the Earth Europe. (2012, February 8). 'Sustainable palm oil' advert false, says watchdog. Accessed from https://friendsoftheearth.eu/press-release/sustainable-palm-oil-advert-false-says-watchdog/

Froehlich, J. (2009). Promoting energy efficient behaviors in home through feedback: The role of human-computer interaction. *Proceedings of the HCIC Workshop, 9*, NJ, 1–11.

GreenPrint. (2021). "Business of Sustainability Index: March 2021". Accessed from https://greenprint.eco/wp-content/uploads/2021/03/GreenPrint-Business-of-Sustainability-Index_3.2021.pdf

Grillo, N., Tokarczyk, J., & Hansen, F. (2008). Green advertising developments in the US forest sector: A follow up. *Forest Products Journal, 58*(5), 40–46.

Han, N. R., Baek, T. H., Yoon, S., & Kim, Y. (2019). Is that coffee mug smiling at me? How anthropomorphism impacts the effectiveness of desirability vs. feasibility appeals in sustainability advertising. *Journal of Retailing and Consumer Services, 51*, 352–361.

Hartmann, P., Apaolaza Iba´n˜ez, V. & Forcada Sainz, F.J. (2005). Green branding effects on attitude: functional versus emotional positioning strategies. *Marketing Intelligence & Planning, 23*(1), 9–29.

Hartmann, P. & Apaolaza-Iba´n˜ez, V. (2009). Green advertising revisited: conditioning virtual nature experiences. *International Journal of Advertising, 28*(4), 715–739.

Haytko, D. L., & Matulich, E. (2008). Green advertising and environmentally responsible consumer behaviours: Linkages examined. *Journal of Management and Marketing Research*, 7(1), 2–11.

Kangun, N., Carlson, L., & Grove, S. J. (1991). "Environmental advertising claims: A preliminary examination". *Journal of Public Policy & Marketing, 10*(2), 47–58.

Kapoor, P. S., Balaji, M. S., & Jiang, Y. (2021). Effectiveness of sustainability communication on social media: Role of message appeal and message source. *International Journal of Contemporary Hospitality Management, 33*(3), 949–972.

Kim, Y., Baek, T. H., Yoon, S., Oh, S., & Choi, Y. K. (2017). Assertive environmental advertising and reactance: Differences between South Koreans and Americans. *Journal of Advertising, 46*(4), 550–564.

Kirchgeorg, M., & Winn, M. I. (2006). Sustainability marketing for the poorest of the poor. *Business Strategy and the Environment, 15*, 171–184.

Knight, M. (2008). It's not easy being green. CNN. Accessed from www.cnn.com/2008/TECH/science/07/16/greenwash/index.html

Leonidou, L. C., Leonidou, C. N., Palihawadana, D., & Hultman, M. (2011). Evaluating the green advertising practices of international firms: A trend analysis. *International Marketing Review, 28*(1), 6–33.

Lung, S. (2010). Green consumerism: The way to effectively differentiate your products in Asia-Pacific market. Accessed from https://ezinearticles.com/?Green-Consumerism---The-Way-to-Effectively-Differentiate-Your-Products-in-Asia-Pacific-Market&id=4875312

Marsden, P., & Chaney, P. (2013). *The social commerce handbook: 20 Secrets for turning social media into social sales*. New York: McGraw Hill.

Matthes, J., Wonneberger, A., & Schmuck, D. (2014). Consumers' green involvement and the persuasive effects of emotional versus functional ads. *Journal of Business Research, 67*(2014), 1885–1893.

Mayer, J. M., Kumar, P., & Yoon, H. J. (2019). Does sexual humor work on Mars, but not on Venus? An exploration of consumer acceptance of sexually humorous advertising. *International Journal of Advertising, 38*(7), 1000–1024.

Minton, E., Lee, C., Orth, U., Kim, C., & Kahle, L. (2012). Sustainable marketing and social media: A cross-country analysis of motives for sustainable behaviours. *Journal of Advertising, 41*(4), 69–84.

Mohd Helmi, A. R., Ros Zayanah, J. A. Z., Fauziah, A., & Lyndon, N. (2012). Green advertising and environmentally responsible behavior: The level of awareness and perception of Malaysian youth. *Asian Social Science, 8*(5), 46–54.

MPOC. (2009). Malaysian palm oil council disagrees with british advertising authority judgment. Accessed from https://mpoc.org.my/malaysian-palm-oil-council-disagrees-with-british-advertising-authority-judgment/

Murphy, R., Graber, M., & Stewart, A. (2010). Green marketing: A study of the impact of green marketing on consumer behavior in a period of recession. *Business Review, 16*(1), 134–140.

Nihon Keizai Newspaper, Advertising Section. (1994). *Read business from newspaper advertising.* Tokyo: Nihon Kezai Newspaper Publishing.

Norazah, M. S. (2013). Green awareness effects on consumers' purchasing decision: Some insights from Malaysia. *International Journal of Asia Pacific Studies, 9*(2), 49–63.

Ongkrutraksa, W. (2002). A content analysis of environmental advertising: Studies of Japanese, Thai, Chinese and American advertisements. In Ramazwick & Tu Ping (Eds.), *AP-Asia Pacific advances in consumer research* (Vol. 5, pp. 120–126). Valdosta, GA: Association for Consumer Research.

Ottman, J. A. (1993). *Green marketing: Challenges and opportunities.* Chicago: NTC Business Books.

Paco, A. M. F. D., & Reis, R. (2012). Factors affecting skepticism toward green advertising. *Journal of Advertising, 41*(4), 147–155.

Parker, J. (2021). Transforming advertising practices for a sustainable future. Accessed from https://www.businesstoday.com.my/2022/02/19/transforming-advertising-practices-for-a-sustainable-future/

Pavan, M. P. S. (2010). Green marketing in India: Emerging opportunities and challenges. *Journal of Engineering, Science and Management Education, 3*, 9–14.

Pittman, M., Oeldorf-Hirsch, A., & Brannan, A. (2022). Green advertising on social media: Brand authenticity mediates the effect of different appeals on purchase intent and digital engagement. *Journal of Current Issues & Research in Advertising, 43*(1), 106–121.

Pulizzi, J. (2014). *Epic content marketing: How to tell a different story, Break through the clutter and win more customers by marketing less.* New York: McGraw Hill.

Randle, M., Kemperman, A., & Dolnicar, S. (2019). Making cause-related corporate social responsibility (CSR) count in holiday accommodation choice. *Tourism Management, 75*, 66–77.

Reilly, A. H., & Hynan, K. A. (2014). Corporate communication, sustainability and social media: It's not easy (really) being green. *Business Horizons, 57*, 747–758.

Saereerat, S., (1995). *Marketing for environment.* Bangkok: Pattanasueksa Publishing.

Sathiswaran, U., Azlinda Shazneem, M. S., & Nurul Hasliana, H. (2014). The role of green marketing towards purchase intention among Malaysian customers. *Journal of Entrepreneurship and Business, 2*(1), 1–8.

Segev, S., Fernandes, J., & Hong, C. (2016). Is your product really green? A content analysis to reassess green advertising. *Journal of Advertising, 45*(1), 85–93.

Shaw, M. (2009). *Copywriting: Successful writing for design, advertising and marketing.* London: Laurence King.

Shen, Y., Lever, M., & Joppe, M. (2020). Investigating the appeal of a visitor guide: A triangulated approach. *International Journal of Contemporary Hospitality Management, 32*(4), 1539–1562.

Shrum, L., McCarty, J., & Lowrey, T. (1995). Buyer characteristics of the green consumer and their implications for advertising strategy. *Journal of Advertising, 24*(2), 71–82.

Story, J. (2019). What is sustainable marketing and how should you use it? *Smart Insights.* Accessed from https://www.smartinsights.com/online-brand-strategy/brand-positioning/sustainable-marketing-how-should-you-use-it/

Tolkes, C. (2020). Sustainability Communication in tourism-a literature review. *Tourism Management Perspectives, 27*, 10–21.

Van den Ende, J., Frederiksen, L., Prencipe, A. (2014). The front end of innovation: Organizing search for ideas. *Journal Product Innovation Management, 32*(4), 482–487.

Wang, S., & Lehto, X. (2019), "The interplay of travelers' psychological distance, language abstraction, and message appeal type in social media advertising", *Journal of Travel Research, 59*(8), 1430–1446.

We Are Social. (2022). Digital 2022 global overview report: The essential guide to the world's connected behaviors. Accessed from https://wearesocial.com/uk/blog/2022/01/digital-2022-another-year-of-bumper-growth-2/

World Wild Life. (2021). Search for sustainable goods grows by 71% as 'Eco-Wakening' grips the globe. Accessed from https://www.worldwildlife.org/press-releases/search-for-sustainable-goods-grows-by-71-as-eco-wakening-grips-the-globe

Ying, S., & Shanyong, W. (2020). Understanding consumers' intentions to purchase green products in the social media marketing context. *Asia Pacific Journal of Marketing and Logistics, 32*(4), 860–878.

Yu, X. (2014). Is environment 'a city thing' in China? Rural–urban differences in environmental attitudes. *Journal of Environmental Psychology, 38*, 39–48.

Zhu, Q., & Sarkis, J. (2016). Green marketing and consumerism as social change in China: Analyzing the literature. *International Journal Production Economics, 181*, 289–302.

Zinkhan, G. M., & Carlson, L. (1995). Green advertising and the reluctant consumer. *Journal of Advertising, 24*(2), 1–5.

7

INTO THE PRACTICES OF MALAYSIAN GAMING INFLUENCERS

Do They Promote Sustainable Development?

Nurzali Ismail

Introduction: Youth and Online Gaming

Digital practices have become an integral part of every facet of youth's lives (OECD, 2018). Malaysian youth, in general, spend long hours online performing various digital practices, such as using social media and playing mobile games (Ismail, 2020). It is through their everyday digital practices that youth build and maintain relationships, learn, work, share repertoires, and negotiate identities (Ismail, 2016; 2020).

One of the main digital practices that Malaysian youth regularly perform is online gaming (Kemp, 2021). Increased connectivity and mobile device ownership among youth in Malaysia have positively influenced the growth of esports and gaming in the country (Kemp, 2021; Nurhayati-Wolff, 2020). In 2019, Malaysians, particularly youth, spent over US$640 million on gaming, making Malaysia one of the market leaders in the region (Mohtar & Si, 2020; Low, 2021).

Besides playing games online, youth also enjoy watching live-streamed esports and gaming content (YPulse, 2021). It was reported that at least one in every five young people watches esports content, and almost half of them follow professional esports competitions (YPulse, 2021). According to Taylor (2020), gaming is a spectator activity that provides pleasure to audience. Youth watch esports content to learn different aspects of gaming, for escapism and entertainment purposes, and because they like esports personalities or influencers (Hamari & Sjoblom, 2017; Taylor, 2020).

The demand for esports and gaming entertainment content led to the emergence of influencers who are known for their ability to attract a huge audience (Geyser, 2021). Gaming influencers such as PewDiePie, Markiplier, and Ninja have combined hundreds of millions of followers on social media (Mediakix, 2019). According to Lou and Kim (2019), youth regard gaming influencers as

DOI: 10.4324/9781003309642-7

being very close and personal to them, owing to the reciprocal relationships that are built and have evolved over time on social media. This enables gaming influencers to influence youth's attitude, an attribute that is appealing to marketers (Boerman & Van Reijmersdal, 2020; Geyser, 2021).

As opinion leaders and attractions to the younger generations (Simic, 2021), gaming influencers have the potential to become agents of social change. This is beyond the role that they are often associated with in marketing and branding (Lepper, 2021). For instance, gaming influencers can effectively do their part in educating youth and promoting good causes such as charity and sustainable development (Daniels, 2022).

Similarly, prominent Malaysian gaming influencers are able to exert a strong influence on youth in the country. This takes into account that, as of 2021, the country had over 20 million active gamers (Marketing Interactive, 2021). While online streaming provides a platform for gaming influencers to showcase their content and earn economic rewards (Miachon, 2018), an important question arises: do they use their fame and influence to promote the sustainable development agenda?

This chapter aims to unravel gaming influencers' practices and their involvement in promoting sustainability within the Malaysian context. This entails the development of esports and gaming in Malaysia, their ecosystem, influencers' digital practices, and their contribution to sustainability. In this chapter, the terms 'esports' and 'gaming' are both used. While the philosophical definitions of game-related terminologies are varied, complicated, and confusing (Arjoranta, 2019), in this chapter, the author uses 'esports' to refer to competitive gaming, whereas 'gaming' refers to playing video games in a broader sense, including for leisure purposes (Banyai et al., 2020).

The Development of Esports and Gaming in Malaysia

Esports and gaming in Malaysia have made significant progress in recent years. Since 2018, Malaysia has recognised the status of esports, putting it on par with traditional sports (Lee, 2018), despite ongoing contention over whether esports should be regarded as sports (Leis et al., 2021). This was followed by the inception of the Strategic Plan for Esports Development 2020–2025 blueprint and other related initiatives, which indicate the government's commitment to sustain the growth of esports in the country (Ashton, 2019; Sengupta, 2020). The long-term aim is to make Malaysia the leading esports hub in Asia (*New Straits Times*, 2018).

The growth of Malaysia's esports and gaming industry appears to be on the right track (Ismail, 2021), considering the upward global trend in the market, revenue, and advertising (Insider Intelligence, 2021). In the 2022 budget, a US$4.78 million allocation was provided by the government to spur the development of esports (Jacob, 2021b). The achievements of homegrown esports athletes were also mentioned in the 2022 budget speech, which indicates the rising status of esports in Malaysia (*The Star*, 2021b).

As of 2020, Malaysia had over 20 million gamers, who play games for leisure on multiple platforms and consoles, and over 24,000 active esports players (Marketing Interactive, 2021; Hirschmann, 2021). *Dota 2* was found to be the most popular esports game in Malaysia, where the country was represented by 36 professional teams and the combined prize money was over US$6.7 million (Hirschmann, 2021; eSportsflag, n.d.). Other popular esports and gaming titles in Malaysia include *Player's Unknown Battlegrounds Mobile* (PUBG), EA Sports' *FIFA* series, and *Mobile Legends: Bang Bang* (MLBB; Tariq, 2021). Over the years, Malaysia has organised many esports tournaments at various levels, including the Malaysia Esports League 2021 and the internationally renowned ESL One tournament and the Kuala Lumpur Major (Tariq, 2021; Ring, 2017).

Despite the encouraging development of esports and gaming in the country, it is worth noting that there is a long way to go. The esports and gaming industry in Malaysia is still in its infancy, particularly its ecosystem, which includes how it should be organised, positioned, and marketed within the mainstream market (Toh, 2019). Without a proper ecosystem, it would be a struggle for the Malaysian esports and gaming industry to reach greater heights, especially with regard to its governance structure. As argued by Peng et al. (2020), without a good governance structure, the industry is disorganised and faces difficulty sustaining growth.

There are roadblocks that the country's esports and gaming fraternity needs to overcome to ensure that the industry can progress further. One of the main obstacles is the negative perception of esports and gaming in Malaysia (Lai, 2019; Jacob, 2021a). As is the case elsewhere, issues such as toxic culture, cheating, gambling, gender discrimination, and drug addiction in esports and gaming (Freitas et al., 2019; Cote, 2020; Gardner, 2021) are also experienced in Malaysia (Yeung, 2020; Yusoff & Yunus, 2021; Tariq, 2021).

Recently, a local esports team and a number of players involved were banned from participating in a national-level tournament as a result of cheating (eGG Network, 2022). There have also been incidents in the past where local esports tournament organisers failed to deliver the promised prize money to the winners, which indicates a lack of professionalism and unethical behaviour (Xiung, 2018).

Although Scholz (2019) posited that the main focus of the esports and gaming industry is on the economic dimension owing to its profit-making aims, it should be understood that a contribution to society is essential too. For instance, during the December 2021 floods that severely hit several Malaysian states, a group of local esports personalities and gaming content creators came together to help with the relief work (Wong, 2021). Involvement with good causes can go a long way to improving public perceptions of esports and gaming in the country.

Gamers as Influencers

Esports and gaming stakeholders are diverse, comprising game publishers, esports teams, players, gamers, fans, sponsors, suppliers, and investors, among others (Miachon, 2018; Scholz, 2019; Scholz, 2020). In this particular section, the author focuses mainly on the practices of gaming influencers with huge followings on social media.

Whether or not gamers decide to play competitively as esports players, to make gaming as a career, or to simply play for leisure depends largely on their own motivations (Banyai et al., 2020). Nonetheless, over the last few years, many gamers have decided to embark on a career as a content creator – someone who showcases gaming content on live-streaming platforms. Besides earning income from sources such as esports competition prize money and team sponsorship, they can get paid for showcasing their online gaming content live through various means such as fans' subscriptions, donations, advertisements, marketing, branding, merchandise sales, and access to exclusive content (Tenbarge, 2019; Geyser, 2020).

Gamers can establish themselves at a higher level by forming a strategic partnership with streaming platforms such as Facebook, Twitch and YouTube (Mazique, 2019). The partnership provides avenues for gamers to attract more followers and to monetise gaming. It is through this path that many gamers develop their careers and eventually become influencers (Michaeloudis, 2021; Minor, 2021).

While there is a huge opportunity for gamers to thrive as influencers owing to the high levels of digital media consumption and keen interest, particularly among youth (Wielki, 2020; YouGov, 2021), it is important for them to find their niche and identity in order to make themselves stand out from the rest. In order to succeed, gamers need to work consistently on nurturing their own brands and, at the same time, increase their fanbase (D'Anastasio, 2021).

The Malaysian situation is similar to elsewhere, as the demand for live gaming content is on the rise. The popularity of gaming influencers and the penetration of gaming followers in the Asian region are very high (YouGov, 2021). It is important to note that gaming influencers are the type of influencer most followed by youth in many countries (YouGov, 2021).

Unlike other places where Twitch is regarded as the main platform for gamers to live-stream (Woodcock & Johnson, 2019), many Malaysian gaming influencers prefer to utilise Facebook and YouTube instead. This is probably owing to the difficulty of getting viewers on Twitch (Lim, 2021). However, the situation with regard to gamers' choice of platform may change in the future, depending very much on audience likes and preferences (Lam & Thean, 2018).

Taking into account social media influencer types (Park et al., 2021), some of the Malaysian gaming influencers have managed to reach 'mega influencer' status, with over 1 million followers on social media. Leading Malaysian gaming influencers such as Soloz, Penjahat Soldier, and Fredo Sameon had a combined total of over 5.5 million followers on Facebook alone, and they also make their presence felt on other social media platforms, including YouTube, Instagram, and TikTok.

Many active Malaysian gaming influencers perform live-stream sessions regularly, with some of them managing to attract over 100,000 viewers consistently for every live session conducted. The reach and engagement rate that they achieve, which can be measured through interactions such as likes, shares, and comments, are also high (Terng, 2020a, 2020b).

In terms of strategies, the more experienced Malaysian gaming influencers are able to distinguish themselves by adopting different styles. This is evident in the choices of game titles they play, tips sharing, interaction, and other online activities. The key strategy here is their ability to effectively engage the audience members, who watch the live-stream gaming session for various purposes including learning, entertainment, relaxation, and as a way to cope with difficult life challenges (De Wit et al., 2020). Gaming influencers' ability to engage the audience through immersion, immediacy, interaction, and sociality derives from their knowledge, experience, and skills in curating content (Haimson & Tang, 2017).

The Role of Gaming Influencers in Promoting Sustainability Causes

Beyond the impact of gaming influencers in marketing, which is becoming more apparent (Haenlein et al., 2020; Aprialita & Sari, 2020; James, 2020; Haughey, 2020), it is worth noting that they also have the capability to influence audience attitudes, particularly youth attitudes towards good causes (Chwialkowska, 2019; Bonnevie et al., 2020). According to Mithika (2022), gaming publishers, influencers, and event organisers have conducted many successful wildlife and humanitarian awareness campaigns, as well as fundraising events in the past, contributing to the well-being of society.

One of the most important causes that gaming influencers should embark on is to ensure people's well-being through a sustainability agenda that emphasises balanced development of the economy, society, and the environment (Zidansek, 2007). Over the years, countries such as Sweden, Norway, Germany, Canada, and Japan, have shown great interest in sustainable development (Berry, 2021). Many initiatives have been implemented to support the sustainability agenda (Andrejevaite, 2021). Similarly, increased awareness about sustainability has also been reported among developing countries such as Malaysia, India, and Thailand (Phoorisart, 2017; Alves, 2020; Lee, 2021). Youth, in general, are keen to do their part to address the global sustainability challenges (United Nations, 2019). Their major concerns include climate change, pollution, land conservation, animal rights, and nuclear proliferation (Patel, 2017).

Despite increased awareness in recent years, continuous efforts are necessary to continually educate youth and empower them to participate actively in the sustainable development agenda (Gill et al., 2019; Ogamba, 2019). Besides the Gaming for Good strategy, which is successful in connecting youth with various stakeholders (Myers & Murabit, 2019; UNEP, 2020; ITU News, 2021), gaming influencers can also play a more proactive role in promoting sustainable development to youth (Patterson & Barratt, 2019). They should take up the responsibility as role models to educate the young audience on responsible development and to influence youth's behaviour towards sustainability. Figure 7.1 illustrates the framework for collaboration with influencers to promote sustainable development.

While making use of esports and gaming to promote sustainability development initiatives can be controversial owing to issues such as excessive electricity usage and screen time, as well as the environmental impact involved in the production of gaming consoles (Patterson & Barratt, 2019; Gordon, 2019, 2020), the potential to reach a mass audience, particularly youth, should not be underestimated (YouGov, 2021). For instance, in 2021, a *Fortnite* esports tournament was organised featuring gaming influencers including Henrik McLean, Gee Nelly, and TN25, with the aim of promoting environmental sustainability, science, and technology (Abbasi, 2021). The event became the first carbon-neutral *Fortnite* tournament, which resulted in 557 trees being planted (Miller, 2021; Lepper, 2021).

In Malaysia, promoting sustainable development is not yet a common practice among gaming influencers. But this does not mean that Malaysian gaming influencers do not support good causes. Over the years, Malaysian gaming influencers have performed many fundraising initiatives and participated in community service activities to achieve various goals and purposes (Sidhu, 2018; Wong, 2021; *The Star*, 2021a). For instance, fan favourite and leading Malaysian gaming influencer Soloz frequently organises his own charity initiatives during live-stream sessions. In an effort to ease the difficulties faced during the pandemic, Soloz and other gaming influencers such as Master Ramen and Sync participated in a Mobile Legends show match to contribute to a COVID-19 fundraising drive (Malaysiakini, 2020). Charity gaming events that are conducted in collaboration with businesses, game publishers,

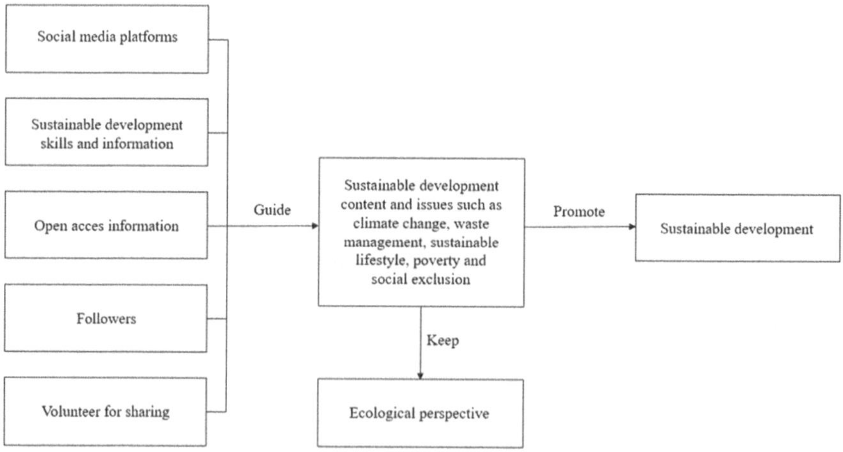

FIGURE 7.1 Framework for collaboration with influencers to promote sustainable development.

and gaming influencers often generate huge interest among fans and provide a good opportunity to raise funds (Ahmad, 2020).

Another Malaysian gaming influencer, Sir Cloud, initiated a fundraising project to save stray cats in his live-stream sessions. He also participated in animal welfare initiatives that were organised by non-governmental organisations (Hayate, 2021). Besides fundraising, other Malaysian gaming influencers also contribute to society in different ways. Take, for example, OnTheGo, who conducted free online intensive additional mathematics classes for youth who were preparing for the national examinations. These efforts indicate gaming influencers' positive attitude and willingness to contribute to society's well-being. More importantly, the good deeds could have a positive impact on young fans, inspiring them to take up similar initiatives like their idols.

There is no doubt that many Malaysian gaming influencers have good intentions in using their gaming practices and reputation for good causes (Malaysiakini, 2020; Wong, 2021; Hayate, 2021). However, in the context of sustainable development, they may not have the necessary awareness required to promote it. Gaming influencers need the authority to successfully engage the audience on sustainable development (Palazzo et al., 2021). In this regard, authority or credibility refers to gaming influencers' ability to exert their influence through either 'authentic' (real) authority or 'apparent' (not real) authority (Garwol, 2020). Here, authentic authority refers to one's ability to demonstrate expertise and substantial knowledge in an area, whereas apparent authority signifies one's lack of expertise and competence, which may possibly lead to misinformation (Garwol, 2020).

Despite their ability to influence attitudes and behaviour, it is not easy for gaming influencers to promote sustainability initiatives on their own. As pointed out by Jajali and Khalid (2021), credibility is a very important consideration in determining whether an influencer is able to influence the audience's attitude with regard to sustainability issues.

It should be understood that, in promoting sustainable development, gaming influencers are considered third-party endorsers who can help to effectively influence youth's attitudes and perceptions on social media (Okuah et al., 2019). Hence, collaboration with stakeholders such as government agencies, businesses, and civil societies is necessary. This can be performed through strategies including direct engagement, tagging and hashtags, emojis, timing, content curation, and incentives (Okuah et al., 2019). Collaboration efforts could also help gaming influencers to overcome potential issues arising owing to lack of authority.

Yildirim (2021) provided a framework for collaboration with influencers that can be adopted in future initiatives for the promotion of sustainable development involving gaming influencers and other stakeholders. The framework comprises social media influencers' key tools for promoting sustainable development – social media platforms, sustainable development skills and information, open access information, followers, and volunteer for sharing (Yildirim, 2021). This is shown in Figure 7.1.

In the long term, it is also proposed that relevant stakeholders such as the government, corporations, and non-governmental organisations come up with strategic collaborations with esports governing bodies and teams, with the aim to educate young gaming enthusiasts on various aspects of sustainable development. This can be done in the form of education module development, as well as through promotion initiatives during events and competitions. Sustainable development should be understood from a more holistic perspective. It is not limited to environmental concerns alone, but also involves other issues including community development, poverty, peace, and prosperity. The Global Esports Federation's strategic collaboration with the United Nations Educational, Scientific and Cultural Organization (UNESCO) to promote the sustainable agenda to youth (Pavitt, 2021) is an example of how such cooperation can be effectively forged within the Malaysian context.

Conclusion

Malaysian youth have a strong interest in esports and gaming. Besides playing games themselves, Malaysian youth also like to watch live gaming content to learn new tips and connect with their esports and gaming idols. The demand for live gaming content has led to the rise of gaming influencers who are capable of influencing young audiences. Some Malaysian gaming influencers have even reached 'mega influencer' status owing to their ability to attract huge numbers of followers and high engagement online.

Considering the importance of gaming influencers in the lives of youth, they should be included in initiatives to promote good causes to youth, including sustainable development. It is worth noting that youth are technologically sophisticated, like entertainment, and are concerned about sustainability issues (Mico-Sanz et al., 2020).

Malaysian gaming influencers, in general, are keen to contribute towards the society's well-being. This is evident in fundraising and other community service-related activities in which they have participated in the past. However, they may not have the awareness required to contribute to the promotion of sustainable development. Bearing in mind their influence and impact as third-party endorsers, relevant stakeholders should strategically collaborate with gaming influencers in the future to ensure the success of sustainable development promotion campaigns and activities in Malaysia.

References

Abbasi, G. (2021). Fortnite's 'gaming for good' tournament shines a spotlight on sustainability. *The Standard*. Accessed from https://www.standard.co.uk/optimist/sustainable/fortnite-gaming-for-good-tournament-sustainability-b952686.html

Ahmad, M. (2020). Team Moon kalahkan Team Soloz dalam Maggi charity showdown. *WTM Today*. Accessed from https://www.wtmtoday.my/team-moon-kalahkan-team-soloz-dalam-maggi-charity-showdown/

Alves, G. (2020). Majority Indians now purchase items based on social responsibility, inclusiveness and environmental impact. *The Economic Times*. Accessed from https://economictimes.indiatimes.com/magazines/panache/majority-indians-purchasing-based-on-social-responsibility-inclusiveness-and-environmental-impact/articleshow/77299600.cms?from=mdr

Andrejevaite, G. (2021). Top 10 sustainable initiatives. *Sustainability*. Accessed from https://sustainabilitymag.com/top10/top-10-sustainable-initiatives

Aprialita, S. D., & Sari, D. M. F. P. (2020). The influence of content, influencer, and experiential marketing on customer engagement of online game applications. *Review of Management, Accounting and Business Studies, 1*(1), 36–48.

Arjoranta, J. (2019). How to define games and why we need to. *The Computer Games Journal, 8*, 109–120.

Ashton, G. (2019, December 12). Malaysia's government lays out five-year strategic plan for esports development. *The Esports Observer*. Accessed from https://archive.esportsobserver.com/malaysia-5year-esports-plan/

Banyai, F., Zsila, A., Griffiths, M. D., Demetrovics, Z., & Kiraly, O. (2020). Career as a professional gamer: Gaming motives as predictors of career plans to become a professional esport player. *Frontiers in Psychology, 11*, 1–9.

Berry, I. (2021). Top 10 greenest countries. *Sustainability*. Accessed from https://sustainabilitymag.com/top10/top-10-greenest-countries-2

Boerman, S. C., & Reijmersdal, E. A. (2020). Disclosing influencer marketing on YouTube to children: The moderating role of para-social relationship. *Frontiers in Psychology, 10*, 1–15.

Bonnevie, E., Rosenberg, S. D., Kummeth, C., Goldbarg, J., Wartella, E., & Smyser, J. (2020). Using social media influencers to increase knowledge and positive attitudes toward the flu vaccine. *PLoS ONE, 15*(10), 1–14.

Chwialkowska, A. (2019). How sustainability influencers drive green lifestyle adoption on social media: The process of green lifestyle adoption explained through the lenses of the minority influence model and social learning theory. *Management of Sustainable Development Sibiu, 11*(1), 33–42.

Cote, A. (2020). Here's what it'll take to clean up esports' toxic culture. *Yahoo News*. Accessed from https://malaysia.news.yahoo.com/heres-itll-clean-esports-toxic-123810209.html

D'Anastasio, C. (2021). Gaming influencers are the future of esports. *Wired*. Accessed from https://www.wired.com/story/gaming-influencers-future-of-esports/

Daniels, T. (2022). How Galaxy Racer content creator AboFlah raised $11m and broke two Guinness World Records. *Esports Insider*. Accessed from https://esportsinsider.com/2022/02/galaxy-racer-aboflah/

De Wit, J., Der Kraan, A. V., & Theeuwes, J. (2020). Live streams on Twitch help viewers cope with difficult periods in life. *Frontiers in Psychology, 11*, 1–16.

eGG Network. (2022). *MPLMYS9: MAB Esports penuhi slot terakhir musim liga*. Accessed from https://egg.network/bahasa/mplmys9-mab-esport-penuhi-slot-terakhir-musim-liga/

eSportsflag. (n.d.). *Most played esports games in Malaysia*. Accessed from https://esportsflag.com/malaysia

Freitas, B. D. A., Contreras-Espinosa, R. S., & Correia, P. A. P. (2019). The benefits and risks of sponsoring esports: A brief literature review. In J. Hiltscher & T. M. Scholz (Eds.), *Esports yearbook 2017/18* (pp. 49–57). Norderstedt, Germany: BoD-Book on Demand.

Gardner, M. (2021). The ugly hatred faced by women in esports, and how we fight back. *Forbes*. Accessed from https://www.forbes.com/sites/mattgardner1/2021/03

/19/the-ugly-hatred-faced-by-women-in-esports-and-how-we-fight-back/?sh=5cf6aca437d8

Garwol, K. (2020). Influencers-contemporary authorities of the young generation? *European Journal of Sustainable Development, 9*(4), 273–280.

Geyser, W. (2020). The ultimate guide to becoming a gaming influencer on livestream platforms. *Influencer Marketing Hub.* Accessed from https://influencermarketinghub.com/how-to-be-gaming-influencer/

Geyser, W. (2021). The incredible growth of esports [+ esports statistics]. *Influencer Marketing Hub.* Accessed from https://influencermarketinghub.com/esports-stats/

Gill, S. A., Aftab, R., Rehman, S. U., Javaid, S. (2019). Youth empowerment and sustainable development: An evidence from Pakistan's Prime Minister's youth program. *Journal of Economic and Administrative Sciences, 35*(3), 202–219.

Gordon, L. (2019). The environmental impact of a PlayStation 4. *The Verge.* Accessed from https://www.theverge.com/2019/12/5/20985330/ps4-sony-playstation-environmental-impact-carbon-footprint-manufacturing-25-anniversary

Gordon, L. (2020). The many ways video game development impacts the climate crisis. *The Verge.* Accessed from https://www.theverge.com/2020/5/5/21243285/video-games-climate-crisis-impact-xbox-playstation-developers

Haenlein, M., Anadol, E., Farnsworth, T., Hugo, H., Hunichen, J., & Welte, D. (2020). Navigating the new era of influencer marketing: How to be successful on Instagram, TikTok, & co. *California Management Review, 63*(1), 5–25.

Haimson, O. L., & Tang, J. C. (2017). What makes live events engaging on Facebook live, Periscope, and Snapchat? In Proceedings of the 2017 CHI Conference on Human Factors in Computing Systems, Denver, USA, 48–60.

Hamari, J., & Sjoblom, M. (2017). What is esports and why do people watch it? *Internet Research, 27*(2), 211–232.

Haughey, C. (2020). The secrets to marketing in the gaming industry. Digital Marketing Institute. Accessed from https://digitalmarketinginstitute.com/blog/how-gaming-influencers-sway-marketing-in-the-gaming-industry

Hayate, M. (2021). Sir Cloud bantu Persatuan Pencegah Penganiayaan Terhadap Haiwan (SCPA). *My Game On.* Accessed from https://www.mygameon.my/news/sir-cloud-bantu-persatuan-pencegah-penganiayaan-terhadap-haiwan-scpa

Hirschmann, R. (2021). Total number of active players in esports in Malaysia from 2016 to 2020. *Statista.* Accessed from https://www.statista.com/statistics/1257532/malaysia-number-of-active-esports-players/

Insider Intelligence. (2021). *Esports ecosystem in 2022: Key industry companies, viewership growth trends, and market revenue stats.* Accessed from https://www.insiderintelligence.com/insights/esports-ecosystem-market-report/

Ismail, N. (2016). Young people's use of new media: Learning through participation in communities of practice. *Malaysian Journal of Communication, 32*(2), 42–64.

Ismail, N. (2020). Youth's usage of new media: Exploring learning and identity formation. *Qualitative Report, 25*(10), 3733–3748.

Ismail, N. (2021). Malaysia needs to leverage growth of esports. *The Malaysian Reserve.* Accessed from https://themalaysianreserve.com/2021/05/28/malaysia-needs-to-leverage-growth-of-esports/

ITU News. (2021, January 27). *Gaming for good: How Playmob connects gamers to the SDGs.* Accessed from https://www.itu.int/hub/2021/01/gaming-for-good-how-playmob-connects-gamers-to-the-sdgs/

Jacob, A. (2021a). Ready to take on the world. *The Sun.* Accessed from https://www.thesundaily.my/spotlight/ready-to-take-on-the-world-DD8261757

Jacob, A. (2021b). Budget 2022 helps to uplift youths: A total 25 initiatives to help give youths a head start in these challenging times. *The Sun*. Accessed from https://www.thesundaily.my/spotlight/budget-2022-helps-to-uplift-youths-AI8596406

Jajali, S. S., & Khalid, H. (2021). The influence of Instagram influencers' activity on green consumption behavior. *Business Management and Strategy, 12*(1), 78–90.

James, D. (2020). Going Mobile: Using influencer marketing to launch a classic game to an all-new audience. *Journal of Digital & Social Media Marketing*, 7(4), 298–305.

Kemp, S. (2021). Digital 2021: Malaysia. We are social. *DataReportal*. Accessed from https://datareportal.com/reports/digital-2021-malaysia

Lai, A. (2019). Public perception on esports hinges on 2019 SEA Games. *IGN Southeast Asia*. Accessed from https://sea.ign.com/levelupkl/155191/news/public-perception-on-esports-hinges-on-2019-sea-games

Lam, J. C., & Thean, R. (2018). An inside look at the harsh reality of streaming in Malaysia. Tilt Report. Accessed from https://www.tiltreport.com/articles/695/inside-look-harsh-reality-streaming-malaysia

Lee, R. (2021). Cover story: Increased awareness of sustainability amid Covid-19. *The Edge*. Accessed from https://www.theedgemarkets.com/article/cover-story-increased-awareness-sustainability-amid-covid19

Lee, S. F. (2018). National Sports Council officially recognizes esports. *Stadium Astro*. Accessed from https://english.stadiumastro.com/sports-others/national-sports-council-officially-recognises-esports-75665

Leis, O., Raue, C., Dreiskamper, D., & Lautenbach, F. (2021). To be or not to be (e) sports? That is not the question! Why and how sport and exercise psychology could research esports. *German Journal of Exercise and Sport Research, 51*, 241–247.

Lepper, J. (2021). Gaming for good: How Fortnite championed sustainability. *Charity Digital*. Accessed from https://charitydigital.org.uk/topics/topics/gaming-for-good-how-fortnite-championed-sustainability-9371

Lim, J. (2021). Local Twitch streamers carve out personal space on internet. *The Edge*. Accessed from https://www.theedgemarkets.com/article/local-twitch-streamers-carve-out-personal-space-internet

Lou, C., & Kim, H. K. (2019). Fancying the new rich and famous? Explicating the roles of influencer content, credibility, and parental mediation in adolescents' parasocial relationship, materialism, and purchase intentions. *Frontiers in Psychology, 10*, 1–17.

Low, M. (2021). The gaming industry in Southeast Asia is levelling up. *New Straits Times*. Accessed from https://www.nst.com.my/opinion/columnists/2021/11/749134/gaming-industry-southeast-asia-levelling

Malaysiakini. (2020). *Mobile legends: Showmatch provides donation to Covid-19*. Accessed from https://www.malaysiakini.com/announcement/526293

Marketing Interactive. (2021). *#MARKiesAwards MY 2021 highlight: Yoodo's bet on gamers drives dominance in esports arena*. Accessed from https://www.marketing-interactive.com/markiesawards-my-2021-highlight-yoodos-bet-on-gamers-drives-dominance-in-esports-arena

Mazique, B. (2019). How to get your gaming channel partnered on YouTube. *Forbes*. Accessed from https://www.forbes.com/sites/brianmazique/2019/01/08/how-to-get-your-gaming-channel-partnered-on-youtube/?sh=e03366461f34

Mediakix. (2019). *30 top gaming influencers to follow*. Accessed from https://mediakix.com/blog/top-gaming-influencers/

Miachon, N. (2018). Esports is the next biggest frontier in influencer marketing. *Forbes*. Accessed from https://www.forbes.com/sites/forbescommunicationscouncil

/2018/06/28/esports-is-the-next-biggest-frontier-in-influencer-marketing/?sh=7a77f49b6d7b

Mico-Sanz, J. L., Diez-Bosch, M., Sabate-Gauxachs, A., & Israel-Turim, V. (2020). Mapping global youth and religion. Big data as lens to envision a sustainable development future. *Tripodos, 48*, 33–52.

Michaeloudis, L. (2021). Gaming influencers: A step-by-step campaign guide. *The Drum*. Accessed from https://www.thedrum.com/industryinsights/2021/05/19/gaming-influencers-step-step-campaign-guide#:~:text=million%20in%202021.-,Gaming%20influencers%20leverage%20live%20video%20content%20to%20drive%20impressive%20engagement,and%20outside%20the%20gaming%20industry.

Miller, G. (2021). Guild esports partners with IET's difference makers campaign to create net-zero Fortnite tournament. *European Gaming*. Accessed from https://europeangaming.eu/portal/latest-news/2021/08/27/98625/guild-esports-partners-with-iets-difference-makers-campaign-to-create-net-zero-fortnite-tournament/

Minor, J. (2021). Twitch and beyond: The best video game live streaming services for 2022. *PCMag Asia*. Accessed from https://sea.pcmag.com/console-games/38449/twitch-and-beyond-the-best-video-game-live-streaming-services

Mithika, B. (2022). What a run: Gaming for a worthy cause. *The Standard*. Accessed from https://www.standardmedia.co.ke/standard-entertainment/article/2001440121/what-a-run-gaming-for-a-worthy-cause

Mohtar, J., & Si, T. M. (2020). The big, future potentials of esports industry as a new driver of economic growth. *Focus Malaysia*. Accessed from https://focusmalaysia.my/the-big-future-potentials-of-esports-industry-as-a-new-driver-of-economic-growth/

Myers, C., & Murabit, A. (2019). Gaming sustainable development. *United Nations*. Accessed from https://www.un.org/sustainabledevelopment/blog/2019/01/gaming-sustainable-development/

New Straits Times. (2018). *Malaysia to be esports hub in Asia: Syed Saddiq*. Accessed from https://www.nst.com.my/sports/others/2018/08/398522/malaysia-be-esports-hub-asia-syed-saddiq

Nurhayati-Wolff, H. (2020). Internet user penetration in Malaysia from 2015 to 2025. Statista. Retrieved from https://www.statista.com/statistics/975058/internet-penetration-rate-in-malaysia/

OECD. (2018). *Children & young people's mental health in the digital age: Shaping the future*. Retrieved from https://www.oecd.org/els/health-systems/Children-and-Young-People-Mental-Health-in-the-Digital-Age.pdf

Ogamba, I. K. (2019). Millennials empowerment: Youth entrepreneurship for sustainable development. *World Journal of Entrepreneurship, Management and Sustainable Development, 15*(3), 267–278.

Okuah, O., Scholtz, B., & Snow, B. (2019). A grounded theory analysis of the techniques used by social media influencers and their potential for influencing the public regarding environmental awareness. In Proceedings of SAICSIT'19, Skukuza, South Africa, 16–18.

Palazzo, M., Vollero, A., Vitale, P., & Siano, A. (2021). Urban and rural destinations on Instagram: Exploring the influencers' role in #sustainabletourism. *Land Use Policy, 100*, 1–11.

Park, J., Lee, J. M., Xiong, V. Y., Septianto, F., & Seo, Y. (2021). David and goliath: When and why micro-influencers are more persuasive than mega-influencers. *Journal of Advertising, 50*(5), 584–602.

Patel, D. (2017). 11 environmental causes gen Z is passionate about. *Forbes*. Accessed from https://www.forbes.com/sites/deeppatel/2017/10/04/11-environmental-causes-gen-z-is-passionate-about/?sh=7b15dbc11849

Patterson, T., & Barratt, S. (2019). *Playing for the planet: How video games can deliver for people and the environment*. Arendal, Norway: UN Environment/GRID-Arendal.

Pavitt, M. (2021). Global esports federation to collaborate with UNESCO on sustainable development goals. *Inside The Games*. Accessed from https://www.insidethegames.biz/articles/1106032/global-esports-federation-unesco-partner

Peng, Q., Dickson, G., Scelles, N., Grix, J., & Brannagan, P. M. (2020). Esports governance: Exploring stakeholder dynamics. *Sustainability, 12*, 1–15.

Phoorisart, N. (2017). Sustainability awareness in Thailand. In T. Issa, P. Isaias, & T. Issa (Eds.), *Sustainability, green IT and education strategies in the twenty-first century. Green energy and technology* (pp. 103–147). Cham: Springer.

Ring, O. (2017). ESL One Dota 2 returns to Genting, Malaysia with $400,000 up for grabs. *Esports Insider*. Accessed from https://esportsinsider.com/2017/10/esl-one-dota-2-returns-genting-malaysia-400000-grabs/

Scholz, T. M. (2019). *Esports is business. Management in the world of competitive gaming.* Cham, Switzerland: Palgrave Macmillan.

Scholz, T. M. (2020). Deciphering the world of esports. *International Journal on Media Management, 22*(1), 1–12.

Sengupta, S. (2020). Malaysian Ministry of youth and sports launches esports integrated. *The Esports Observer*. Accessed from https://archive.esportsobserver.com/malaysia-esports-integrated/

Sidhu, K. K. (2018). Kitamen targets US$510k fundraising via ECF to grow its eSports presence. *Digital News Asia*. Accessed from https://www.digitalnewsasia.com/digital-economy/kitamen-targets-us510k-fundraising-ecf-grow-its-esports-presence

Simic, I. (2021). Gaming influencers: Promotional rockstars of their generation. *Esports Insider*. Accessed from https://esportsinsider.com/2021/11/gaming-influencers-yougov/

Tariq, Q. (2021a). Inaugural Malaysia Esports League launched with RM200000 prize pool. *The Star*. Accessed from https://www.thestar.com.my/tech/tech-news/2021/07/06/inaugural-malaysia-esports-league-launched-with-rm200000-prize-pool

Tariq, Q. (2021b). Ethics in esports. *The Star*. Accessed from https://www.thestar.com.my/tech/tech-news/2021/08/16/ethics-in-esports

Taylor, T. L. (2020). The rise of massive multiplayer online games, esports, and game live streaming: An interview with T. L. Taylor. *American Journal of Play, 12*(2), 107–116.

Tenbarge, K. (2019). Gamers say they're earning more money on Facebook's streaming platform than on Twitch and YouTube. *Business Insider*. Accessed from https://www.businessinsider.com/gamers-earn-more-money-facebook-gaming-twitch-youtube-mixer-streaming-2019-10

Terng, B. (2020a). Soloz's peak viewership reaches new heights. Egg Network. Accessed from https://egg.network/eng/news/solozs-peak-viewership-reaches-new-heights/

Terng, B. (2020b). Facebook gaming: Soloz achieves record-breaking two million followers. Egg Network. Accessed from https://egg.network/eng/news/facebook-gaming-soloz-achieves-record-breaking-two-million-followers/

The Star. (2021a). *Gamer gives back through online platform*. Accessed from https://www.thestar.com.my/metro/metro-news/2021/06/28/gamer-gives-back-through-online-platform

The Star. (2021b). *Budget 2022: RM20mil set aside to develop emerging eSports talents*. Accessed from https://www.thestar.com.my/news/nation/2021/10/29/budget-2022-rm20mil-set-aside-to-develop-emerging-esports-talents

Toh, B. (2019). Can Malaysia's esports go mainstream? *The Edge Markets*. Accessed from https://www.theedgemarkets.com/article/can-malaysias-esports-go-mainstream

UNEP. (2020). Video game industry to rally 250 million players to protect the planet. United Nations Environment Program. Accessed from https://www.unep.org/news-and-stories/press-release/video-game-industry-rally-250-million-players-protect-planet

United Nations. (2019). Youth advocates demand immediate action to address urgent challenges. *Sustainable Development*. Accessed from https://www.un.org/sustainabledevelopment/blog/2019/04/youth-advocates-demand-immediate-action-to-address-urgent-challenges/

Wielki, J. (2020). Analysis of the role of digital influencers and their impact on the functioning of the contemporary on-line promotional system and its sustainable development. *Sustainability*, *12*, 1–20.

Wong, G. (2021). Facebook gaming creators band together to help out flood victims. *eGG Network*. Accessed from https://egg.network/eng/news/facebook-gaming-creators-band-together-to-help-out-flood-victims/

Woodcock, J., & Johnson, M. R. (2019). Live streamers on Twitch.tv as social media influencers: Chances and challenges for strategic communication. *International Journal of Strategic Communication*, *13*(4), 321–335.

Xiung, C. J. (2018). Malaysia's esports scene shows growth but challenges remain. *Digital News Asia*. Accessed from https://www.digitalnewsasia.com/digital-economy/malaysias-e-sports-scene-shows-growth-challenges-remain

Yeung, Y-G. (2020). Gamer 'Tashbunny' opens up on 'toxic' misogyny in Malaysia's gaming industry. *Yahoo News*. Accessed from https://malaysia.news.yahoo.com/gamer-tashbunny-opens-toxic-misogyny-074042266.html?

Yildirim, S. (2021). Do green women influencers spur sustainable consumption patterns? Descriptive evidences from social media influencers. *Ecofeminism and Climate Change*, *2*(4), 198–210.

YouGov. (2021). *Game-changers: The power of gaming influencers*. Accessed from https://commercial.yougov.com/rs/464-VHH-988/images/YouGov_Game-Changers%202021_PART%201.pdf

YPulse. (2021). Here's how many young people would rather watch esports than traditional sports. Accessed from https://www.ypulse.com/article/2021/10/27/heres-how-many-young-people-would-rather-watch-esports-than-traditional-sports/

Yusoff, N. H., & Yunus, Y. H. M. (2021). Male dominant sport: The challenges of esports female athletes. *Pertanika Journal of Social Science and Humanities*, *29*(2), 1415–1429.

Zidansek, A. (2007). Sustainable development and happiness in nations. *Energy*, *32*(6), 891–897.

8

VIRTUAL EXHIBITION OF INDONESIAN GEOPARK AS A SUSTAINABILITY COMMUNICATION EFFORT

Eli Jamilah Mihardja, Togu S. Pardede, and Nurul Selen Azizah ASP

Introduction: Geopark in Indonesia

Geopark is a holistic concept that integrates the conservation and management of three natural forms of diversity, namely geological diversity (geoheritage), biodiversity, and cultural diversity; it aims to develop sustainable areas based on the principle of protection (conservation). In this case, it is carried out collaboratively by the central government, regional governments, and stakeholders across three pillars: conservation, education, and sustainable community economic development.

Geopark development can raise awareness and understanding of the main problems faced by local communities, such as the sustainable use of Earth's resources, mitigating the effects of climate change, and reducing the risk of natural disasters. Geopark development can offer a regional identity and foster a community's sense of pride in the area. In addition, positive impacts will include the creation of innovative local business activities, new job opportunities, and being the locus/theme of development programmes/activities for equitable development.

Geopark development will empower local communities by promoting significant geological processes in the area – features, time periods, geological history, or unique geological beauty – through collaborative activities. The process of founding geoparks is bottom–up, involving all relevant parties, both local and regional (such as landowners, community groups, tourism service providers, indigenous peoples, local organisations, and so on). This process requires commitment from the local community, strong partnerships, political support, and a strategy that will meet all of the community's goals when displaying and protecting the geological heritage of the area.

DOI: 10.4324/9781003309642-8

To support the development of geoparks, various efforts have been made, especially by the Ministry of National Development Planning (PPN)/National Development Planning Agency (Bappenas) as the compilers of the Indonesia Geopark Action Plan. Synergistic efforts continue to be carried out with relevant departments as stakeholders in geopark development. As the conductor, the ministry has created an orchestration plan that is expected to be harmonious, beautiful, and progressive in order to ensure the safety and sustainability of the Earth's heritage. Therefore, efforts are made on a continuous basis to maximise potential and collaborate with various parties in order to raise awareness and engagement with the existence and development of geoparks in Indonesia.

Sustainability Communication

The rise of the term "sustainability communication" has been accompanied by calls for more responsible human interaction with natural and social environments. This requires a social understanding process that addresses potential causes and solutions. The task of sustainability communication is to critically evaluate and introduce an understanding of human–environment relations into social discourse.

Sustainability communication is a relatively new concept. It has its roots in a variety of discourses, including environmental, risk, and science communication. On the one hand, these discourses share a number of characteristics, such as a similar subject focus and the importance of the media. There are, however, significant differences between them in terms of theoretical foundation, political reach, and respective actors. Sustainability communication, according to this study, should be viewed as an integrative method that brings together the key features of various communication perspectives (Adomßent & Godemann, 2011).

According to Cahyandito (2010), the concept of sustainable development has been promoted since the publication of the Brundtland Report in 1980, followed by the Rio de Janeiro Summit Conference on Environment and Development in 1992. One of the potential solutions to this problem is sustainability communication. The term "sustainability communication" was coined in the late 1990s and it has since shifted the emphasis of environmental communication. With this concept, it is hoped that people will learn more about what is going on around them, such as the environmental, social, and economic issues that are impeding the fulfilment of the objective of sustainable development. Communication about sustainability is critical at this time. According to the sustainability communication concept, one of the communication instruments that has gained popularity is sustainability reporting. Producing and distributing a sustainability report can provide an organisation with numerous benefits. Sustainability reporting is now widely regarded as a tool for enhancing a company's image and maintaining a reciprocal connection with stakeholders, as well as being a critical component in successful commercial competitiveness.

In the literature, sustainability communication refers more to the business world. For example, a study by Genç (2017) discusses sustainability communication in business communication activities performed by business organisations. In order to plan and develop sustainability and sustainable strategies, it is necessary for people to express themselves internally and externally using appropriate mode(s) of communication. Thus, in order to establish effective communication, the managers in an organisation have to channel to the receiver what they mean to communicate in a simple, direct, and precise manner, whether using an oral or written modality. To achieve this end, they should also take into consideration gender and cultural variations in terms of communication (Genç, 2017).

Another study refers to sustainability communication on campus (Franz-Balsen & Heinrichs, 2007) and Sustainability Communication in the form of CSR and using new media (Nwagbara & Reid, 2013). Regarding the use of new media, the task of Sustainability Communication is to critically evaluate and introduce an understanding of the human–environment relationship into social discourse (Fischer, Lüdecke, Godemann, Michelsen, Newig, Rieckmann, & Schulz, 2016).

Virtual Exhibition

A virtual exhibition is a type of *virtual* event at which businesses in a particular industry promote their products and services. According to Ciurea, Zamfiroiu, and Grosu (2014), virtual exhibitions provide a number of advantages over traditional exhibitions, such as the ease with which new components can be added to the gallery of objects or current ones can be modified. The ability to have electronic duplicates of real cultural heritage pieces, which are typically very expensive and fragile and can be damaged by handling, is a significant benefit. Mobile applications that take the form of virtual exhibitions and have the goal of presenting rare and valuable objects will see significant development in the future, based on their impact in the educational and cultural spheres. However, we believe that the market for mobile applications is expanding, and that there is a high demand for such products.

Recently, the use of technology to develop the e-business of scenic sites has grown in popularity. Furthermore, a virtual 3D panoramic show is a great way to get the word out. It provides audiences with not only a genuine experience but also an eye-catching visual impression. As a result, this image-based virtual exhibition technology will advance quickly, and more geological parks and other scenic sites will adopt it (Yan, Ma, & Tao, 2009).

Virtual exhibitions have some advantages, such as the ease with which new components can be added or existing ones changed, such as adding new elements, updating and reusing existing ones, or adding new educational components. Visual exhibitions have the added benefit of being able to create and use electronic duplicates of physically fragile records or expensive artefacts, which might be damaged during physical consultation, through digitisation and rendering. Virtual exhibitions, which are implemented as mobile applications and have the goal of presenting rare collections from various cultural domains, such as museums and libraries, will undergo a substantial evolution in the next few years, owing to their educational and cultural influence. New business models for interactions between various players, such as owners of cultural items, physical and virtual visitors, creative firms, service providers and telecom operators, and public bodies, must be adopted for the movement to be viable.

The same is true for geotourism and geoparks. Many geoparks are attempting to generate a wide range of geoproducts for a variety of target audiences. The nature of these products, on the other hand, is highly different, and their evolution can be seen primarily in terms of timescale. Simple geoproduct interpretation and application were once popular, but newer technologies such as animation and augmented reality are now thriving. This is primarily owing to the increased availability of technological equipment. At the same time, a lot of geoparks have exceptional interpreters who can portray and clarify complex geological concepts. They also have the ability to generate new geoproducts where there is a clear need to condense and abbreviate information. The findings of Doucek and Zelenka's (2018) study show a positive trend in the application of current methods to geoscience interpretation and the development of new geoproducts. At the same time, it is apparent that reaching current visitors without a modern concept is quite difficult. As a result, such a concept is required in order to develop geoscientific concerns, particularly in the domain of geoparks.

During the COVID-19 pandemic, the tourism industry started to weaken, especially in Indonesia. In the midst of declining tourism trends, technology is actually growing, including by maximising opportunities for online destination and authenticity, virtual reality, and augmented reality for virtual tourism to cultural heritage sites in Indonesia (Artha, Hadi, & Sari, 2020). Currently, there are many studies on the use of virtual expos for the benefit of tourism marketing and the business world in general. Virtual expos have been used for online exhibitions and museum activities (Budiyanto, Tutuko, Winansih, Setiawan, & Iqbal, 2020; Jaya, 2021; Bramantyo & Ismail, 2021). In addition, virtual expos are also used to market tourist destinations (Satya, Tejaningrum, & Hanifah, 2021; Susanto, Wijonarko, Purwanto, & Binabar, 2019).

Geopark and Sustainable Development Goals (SDGs) in Indonesia

A geopark is a geographical area that has geological heritage and geological diversity of high value, including biodiversity and cultural diversity. Geoparks were founded on three pillars: conservation, education, and local economic development. Six Indonesian geoparks have been designated by the United Nations Educational, Scientific and Cultural Organization (UNESCO) as Global Geoparks, namely Batur Bali, Mount Sewu, Rinjani, Ciletuh, Toba Caldera, and Belitong. There are 15 national geoparks and 110 geological heritage sites that have the potential to become global geoparks.

Geopark development also contributes to Indonesia's efforts to achieve the goals and targets of the Sustainable Development Goals. Geopark development is concretely reflected in eight goals. These range from zero-poverty goals to quality education, decent work, and economic growth, as well as responsible production and consumption. Through the National Geopark I Conference, which was held on 12 July 2018 in the Bappenas Building, the government of Indonesia, through various ministries and institutions, agreed to encourage the development of geoparks to contribute to the SDGs as Indonesia's commitment to sustainable development. There are at least ten goals in the SDGs where geopark development can contribute to their achievement: (1) Goal 1, No Poverty; (2) Goal 4, Quality Education; (3) Goal 5, Gender Equality; (4) Goal 8, Decent Work and Economic Growth; (5) Goal 9: Industry, Innovation and Infrastructure, (6) Goal 11, Sustainable Cities; (7) Goal 12, Responsible Production and Consumption; (8) Goal 13, Climate Action; (9) Goal 14, Live below Water; (10) Goal 15, Life on Land; and (11) Goal 17, Partnership for the Goals.

Geopark development makes real contributions to Indonesia. These contributions include, among other things, regional development, increasing community resilience when facing disasters, and educating people on how to live a good life while respecting diverse cultures. This is in accordance with the concept of sustainable development. Sustainable development is a development process that maximises available natural resources and is processed by humans with development. In general, sustainable development is regarded as a sign of a country's progress in terms of its social, economic, and political structures. In addition, sustainable development can aid in the improvement of people's well-being and the fulfilment of their needs, while not impeding the development of the next generation to do the same.

The main concept of sustainable development is to help improve people's welfare. Furthermore, other key concepts include: (1) managing natural resources well so that they can be used in the future, (2) making the best use of natural resources so that there is no waste and no harm to the environment, (3) development that must be able to increase renewable natural resources, and (4) maintaining the quality of human life in the present and in the future.

The Virtual Indonesia Geopark Summit 2021: Socialising the Significance of Geopark

Various efforts have been made to introduce and socialise geoparks, to the Indonesian people in particular and to the rest of the world in general. The 2021 Indonesia Geopark Summit was a series of activities from 25 October to 2 November. The data described below are part of a report on the 2021 Geopark Summit activities launched by the Geopark National Action Plan.

The series of events included a virtual advisory mission (Figure 8.1). The government's commitment to developing institutional capacity and human resources in the development of geoparks at the regional level, particularly for the Maros-Pangkep National Geopark, Ijen National Geopark, and Aspiring Geopark Gorontalo, is reflected in the virtual advisory mission. In the context of developing a world-class geopark, this activity was attended by all stakeholders, both central and regional. The Maros-Pangkep Aspiring UNESCO Global Geopark virtual advisory mission was held on 26–27 October 2021. Several key points from this virtual advisory mission represent three types of geological distinctiveness. Among the three potentials, the karst area is one of the largest karst mountain clusters in the world and has the most strategic function for the life of the surrounding environment, both for flora and fauna and for humans who inhabit it. It has an extraordinary beauty and is an example of an exotic and rare geological heritage that needs to be preserved and managed for sustainable economic development.

A virtual tour of the natural beauty of Bondowoso and Banyuwangi regencies, which are part of the Crater Ijen-Blue Fire Aspiring UNESCO Global Geopark, was held on 28–29 October 2021. There are 21 geological sites developed in Ijen Aspiring UNESCO Global Geopark, ranging from local to international scales. Among the many geological sites that have extraordinary phenomena are Ijen Crater, the most acidic crater lake on Earth with blue flames caused by Solfatara, and Alas Purwo National Park, which can be seen online and is composed of limestone. Both of these sites exhibit remarkable karst geomorphology.

The Second Geopark Indonesia National Conference (KONAS Geopark II) was held on 22 November 2021 as part of a series of events leading up to the 2021 Geopark Indonesia Summit. KONAS Geopark II was a follow-up to KONAS Geopark I, which took place in 2018. KONAS Geopark Indonesia II was opened by the president of the Republic of Indonesia, Joko Widodo. In his speech, President Jokowi invited all elements of the nation to maintain Indonesia's legacy from God. President Jokowi made several important points, including the fact that geoparks can be developed for education and economic development of the

FIGURE 8.1 Maros-Pangkep and Ijen Aspiring UNESCO Global Geopark virtual advisory missions

local community through geotourism. In addition, stories, narratives, and storytelling in each geopark are areas that need to be strengthened in the development of geoparks.

The coordinating minister for maritime affairs and investment also delivered a keynote speech at the Second Geopark National Conference. The support of seven ministers was also conveyed at the Geopark Summit. According to Minister of National Development Planning Bapak Suharso Monoarfa, Ministerial (PPN) Regulation No. 15 of 2020 concerning the National Action Plan (RAN) for Geopark Development 2021–2025 was enacted to provide a guideline for all stakeholders pursuing sustainable and inclusive implementation of geopark development, as mandated in the Presidential Decree No. 9 of 2019. Geoparks play an important role, as stated by the ministers of education, culture, research, and technology, as well as the minister of tourism and creative economy. Geoparks play an important role as a source of unique knowledge that is characterised by local wisdom and values. Tourism, the creative economy, small medium enterprises (SME), agriculture, culinary, and other sectors can all benefit from the development of geoparks. Geoparks are one of the national tourism strengths that prioritise cultural and environmental preservation.

Geopark development based on conservation management, economic development, and community development, according to the minister of environment and forestry, will benefit nature conservation and support the development of conservation and community-based nature tourism. Collaboration by various stakeholders is required in the development of geoparks. This was communicated by the ministers of energy and mineral resources, as well as state-owned enterprises.

The Geopark Virtual Exhibition ran from 22 November 2021 to 22 January 2022. The Geopark Virtual Exhibition is an exhibition of geopark programmes, geoproducts from MSMEs, and website-based education with the website address geopark.bappenas.go.id. This exhibition was attended by six UNESCO Global Geoparks (UGGs; Batur, Gunung Sewu, Ciletuh-Palabuhanratu, Rinjani-Lombok, Toba Caldera, and Belitong), 13 national geoparks (Maros-Pangkep, Ijen, Raja Ampat, Merangin-Jambi, Karangsambung-Karangbolong, Tambora, Silokek, Sianok-Maninjau, Agam-Sawahlunto, Natuna, Pongkor, Bojonegoro, and Meratus), and also 11 Aspiring Geoparks (Ujung Kulon Pandeglang, Gorontalo, Bayah Dome Lebak, Bromo-Tengger-Semeru, Dieng, Poso Lake, Tulungagung, Krakatau, Sankurilang-Mangkalihat, Rajamandala, and Kelimutu Ende). Visitors to this virtual exhibition can view geopark profiles; photos and videos about existing geoparks in terms of geodiversity, biodiversity, and cultural diversity; and investment profiles for each geopark. In addition to exhibitions from each geopark, there is also a booth for the Indonesian National Geopark Commission. The website is expected to generate investment opportunities and provide investors with an overview of each geopark's general profile. Each booth also has a contact person to help the management agency communicate with potential investors or partners.

The venue of the Geopark Virtual Exhibition is shown in Figure 8.2, and Figure 8.3 demonstrates the booth display.

The Geopark Investment Forum is a forum for meetings between potential investors and the regional government and geopark managers. This forum brings together domestic and foreign investors to identify mutual investment profit opportunities in sustainable development, as well as virtual exhibitions on the introduction of regional potential, economic, social, and environmental programmes, and products and services from local SMEs in the geopark

FIGURE 8.2 Venue in the Geopark Virtual Exhibition.

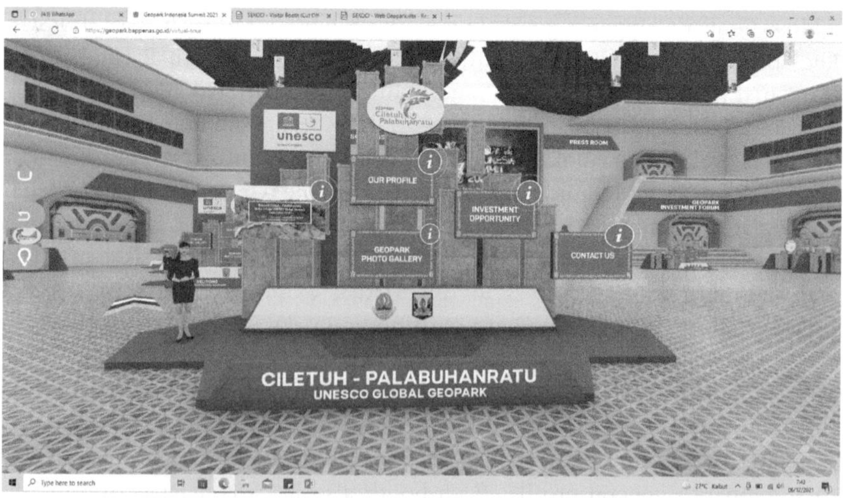

FIGURE 8.3 Booth display in the Geopark Virtual Exhibition.

area. On 23 November 2021, a hybrid activity was held at the Toba Caldera and Bappenas with the theme "Geoparks as Potential Investments in Nature Supporting Sustainable Development Based on Green Economics". The aim of the Geopark Investment Forum is to promote and introduce geoparks as investment destinations for potential investors, local and international. The Geopark Authority may submit high-potential or superior projects for consideration by investors. Through the development of Geopark Indonesia, the meeting became a means of communication and discussion to update on policy directions and investment climate trends in Indonesia and internationally in support of the green economy. This was the first step in introducing geoparks as a potential investment option, and it was to be followed by Geopark EXPO 2022. This expo was to facilitate cooperation between geopark authorities and potential investors, both local and international. It was also meant to introduce and offer geoparks for Indonesian green investment areas to international investors or other international forums.

The minister of investment/BKPM opened the Geopark Investment Forum with the theme "The Role of Investment Support in the Development of Indonesian Geoparks". Green energy and green industry, according to the minister of investment/BKPM, must be categorised as inseparable parts of the concept of developing a tourism economic system based on geoparks. This event was followed by a webinar on the "Role and Collaboration of Stakeholders in Green Geopark Development".

Geoparks originated in Europe, with the goal of protecting geological heritage in European countries through a non-governmental organisation called EGN (Europe Geopark Network) set up in 2001. Furthermore, the world body UNESCO developed and facilitated the existence of geoparks by forming the GGN (Global Geopark Network) organisation in 2004 to accommodate more members from around the world (Du & Girault, 2018). With an increasing number of members, geoparks have evolved beyond simply protecting geological heritage. The goal of a geopark is to take advantage of, explore, appreciate, and develop geological heritage, as well as to preserve biomes (areas or ecosystems that have the same geography and climate). In other words, a geopark is the integration of geological heritage management with an area's cultural heritage for three main reasons: conservation, education, and sustainable development.

The Geological Heritage and Sustainability Tourism

The geopark concept was created on the basis of efforts to build awareness of geological heritage. This geological heritage area can be used to promote the concept of geological heritage preservation and natural resource preservation to the general public, which was UNESCO's initial idea in introducing the geopark concept. Geoparks also aim to keep local communities' economies afloat by promoting geotourism. The geopark concept is regarded as a continuous dynamic and development of the geoconservation movement, which is regarded as the introduction that led to the idea of forming a geopark model (Du & Girault 2018).

Among the eight geological aspects supporting the SDG programme are geological heritage and geotourism (geoheritage and geotourism). Based on a study conducted by the Geological Survey Center team, the development of geological heritage begins with the geodiversity and geoheritage stages, which include inventory, criteria, comparison, and classification activities. Then, in geoconservation, there are activities to protect and preserve the environment. Furthermore, at the end, a geopark will be formed where management, preservation, and utilisation in sustainable development will be realised.

Geodiversity or "global hotspots" are found all over the world, particularly in places with a complex geological history, on continental boundaries (especially convergent boundaries), and with high relief or coastal areas, such as Indonesia. Geological evolution, tectonic processes, rock composition, minerals and fossils, weathering and erosion, and morphology all have an impact on Indonesia's geological heritage. The diversity and uniqueness of the geology provide their own huge potential value for geological heritage and have become the foundation for sustainable development in Indonesia, including the geopark concept. Presidential Regulation No. 9 of 2019 and the technical guidelines issued by the minister of energy and mineral resources, Regulation No. 1 of 2020, regarding the determination of geological heritage sites in the development of Earth parks (geoparks) serve as the foundation for managing geosites in geopark areas.

Sustainable tourism development is inextricably linked to efforts to ensure that the natural, social, and cultural resources used for tourism development in this generation can be enjoyed by future generations. As stated in the Sustainable Tourism Charter of 1995, tourism development must be based on sustainability criteria, which means that development can be ecologically supported in the long term as well as economically feasible and ethically and socially fair to the community. This means that sustainable development is an integrated and organised effort to improve people's lives by regulating the supply, development, use, and maintenance of resources in a sustainable manner. Therefore, a spirit of innovation and tourism creation is required, which must be supported by long-term investment.

Sustainable investing directs investment capital to combat climate change and environmental destruction and preserve the Earth, as well as to promote the responsibility of businesses to act in a sustainable manner, including in the tourism sector, as packaged in the concepts of geopark and astro-tourism as our common heritage and home planet assets. Currently, sustainable investing is the beginning of increasing innovation and creation in national tourism. Tourism is currently one of the largest economic sectors and has the fastest growth rate in the world. Tourism has become one of the main sources of revenue for many countries, including Indonesia. Through foreign exchange earnings, job creation, business opportunities, and infrastructure development, tourism is one of the main drivers of a country's socio-economic progress in 15 main aspects. In Indonesia, one example of such success is UGG Gunungsewu, which has been able to significantly support the community's socio-economic growth, reduce poverty, and increase investment flows in a sustainable manner since its designation as a UNESCO Global Geopark. The Gunungsewu area was previously classified as an underdeveloped area in southern Java that included three provinces: East Java, Central Java, and Yogyakarta.

In recent years, the growth rate of Indonesia's large-scale sustainable tourism sector has reached 25.68%. In the midst of increasingly fierce world tourism competition, innovation, strategy, and productivity are needed to seize the tourism market. Statistics Indonesia (BPS) recorded that, during January–March 2019, the number of foreign tourists reached 3.82 million. In 2018, the number of foreign tourists was recorded at 15.81 million. This is up 12.61% from 2017, which amounted to 14.04 million. This figure is expected to rise further until 2030, in line with the achievement of the Sustainable Development Goals. UGG Gunungsewu is the best example of a UNESCO Global Geopark's role in supporting sustainable national development, with the ability to increase tourist visits by up to 9.1% and regional income by 5.1%, significantly increase investment flows in the tourism sector, and reduce poverty by 4.1%. This information is based on the 2019 Validation Preparation Final Report's résumé.

Sustainability Communication in Geopark Development

Geoparks may be a solution to Indonesia's long-term economic development. The Indonesian government is promoting geoparks as a breakthrough solution for sustainable economic development. The president of the Republic of Indonesia issued Presidential Regulation No. 9 of 2019 concerning the development of Indonesian geoparks, which regulates the existence of the Indonesian National Geopark Committee and the management of Indonesian geoparks. The issuance of this presidential regulation is expected to accelerate the development of geoparks in Indonesia into global geoparks.

Geopark development is a national priority in Indonesia's Mid-Term Development Plan 2021–2024. The presence of geoparks has a significant impact in several areas that have obtained National Geopark and UNESCO Global Geopark status, in fields ranging from education and conservation to sustainable economic growth. One of the most important things in developing a geopark is the realisation of sustainable economic improvement based on local communities. Evidence of these achievements has been seen in several fields, both in areas where the area has become a national geopark and where it is included in the UNESCO Global Geoparks. For example, after Gunung Sewu UGG in Gunung

Kidul Regency was designated as a national geopark and UGG, the increase in regional income was accompanied by a decrease in poverty and an increase in other economic values.

Indonesia is the largest archipelagic country in the world and one of the richest countries in the world in terms of culture and biodiversity. Over 300 ethnic groups are recognised in Indonesia, living alongside each other, as well as over 700 languages. In addition to being located on three major tectonic plates, Indonesia offers valuable geological diversity. All the achievements and potential of this diverse geoheritage site will be less effective without strengthening positioning, differentiation, and branding strategies. Geoparks are a comprehensive and integrated concept for the conservation and utilisation of natural heritage (both geological and biological), as well as cultural wealth for sustainable development that can accelerate economic growth, allowing collaboration of all conservation, education, research, and community development policies.

The development of geoparks in Indonesia helps to achieve the goal of sustainable development. Geopark development can serve as a concrete example of how the government, private sector, and community can collaborate to achieve the Sustainable Development Goals, develop the region holistically and sustainably, and recover and transform the economy following the COVID-19 pandemic toward more sustainable and greener development practices.

Conclusion

Sustainable development is one of the most important concepts if a country wants to continue to develop. However, it should be remembered that sustainable development needs to be carried out by optimising available natural resources and must contribute to improving people's welfare or at least reducing existing gaps. In fact, equity is one of the five principles of sustainable development, the other four being economy, energy, ecology, and engagement or participation. The Sustainable Development Goals set the goal of carrying out sustainable development until 2030, and indicators such as ecological and political sustainability are used to determine the success of sustainable development.

One of the potential solutions to this problem of sustainable development is sustainability communication (Cahyandito, 2010). It is hoped that people will learn more about what is going on around them, such as environmental, social, and economic issues that are impeding the achievement of sustainable development objectives. Therefore, the idea of sustainable development is conveyed through a specific communication strategy, by utilising all existing potential in accordance with the times. One sustainability communication strategy is to use virtual exhibitions. Sustainability communication in the context of sustainable development can be seen in the case of the Geopark Indonesia Virtual Exhibition at the Indonesia Geopark Summit in 2021. Digital space can be reached by virtual exhibitions, and, therefore, the reach for delivering information about geoparks is unlimited. A virtual exhibition has the advantage of being able to include components such as audio-visual items that can enhance the sensory experience. Geoparks are special-interest tourist destinations that are not based on quantity; this opportunity can be used as a pre-experience as well as to screen potential visitors.

References

Adomßent, M., & Godemann, J. (2011). Sustainability communication: An integrative approach. In *Sustainability communication* (pp. 27–37). Dordrecht: Springer.

Artha, B., Hadi, A. S., & Sari, N. P. (2020). Faktor penentu pariwisata virtual situs warisan dunia di Indonesia: Sebuah telaah pustaka dan kerangka konseptual. *Jurnal Kepariwisataan Indonesia: Jurnal Penelitian dan Pengembangan Kepariwisataan Indonesia, 14*(1), 38–45.

Bramantyo, B. D., & Ismail, P. (2021). Digital tourism museum nasional Indonesia melalui virtual tour di masa pandemi COVID-19. *WACANA: Jurnal Ilmiah Ilmu Komunikasi, 20*(2), 184–196.

Budiyanto, H., Tutuko, P., Winansih, E., Setiawan, A. B., & Iqbal, M. (2020). Virtual Expo Menggunakan Panggung sebagai Solusi Pameran Online di Masa Pandemi COVID-19. *Abdimas: Jurnal Pengabdian Masyarakat*, 5(3), 202–208.

Cahyandito, M. F. (2010). Pembangunan berkelanjutan, ekonomi dan ekologi, sustainability communication dan sustainability reporting. *Jurnal Bisnis dan Manajemen, 5*(1), 7–10.

Du, Y., & Girault Y. (2018). A genealogy of UNESCO global geopark: Emergence and evolution. *International Journal of Geoheritage and Parks, 6*(2), 1–17.

Doucek, J., & Zelenka, J. (2018). New trends in geoproducts development: Železné Hory National Geopark case study. *Czech Journal of Tourism*, 7(2), 179–195.

Fischer, D., Lüdecke, G., Godemann, J., Michelsen, G., Newig, J., Rieckmann, M., & Schulz, D. (2016). Sustainability communication. In *Sustainability Science* (pp. 139–148). Dordrecht: Springer.

Franz-Balsen, A., & Heinrichs, H. (2007). Managing sustainability communication on campus: Experiences from Lüneburg. *International Journal of Sustainability in Higher Education, 8*(4), 431–445.

Genç, R. (2017). The importance of communication in sustainability & sustainable strategies. *Procedia Manufacturing, 8*, 511–516.

Jaya, I. K. F. P. (2021, February). Museum Virtual Google Art and Culture Sebagai Media Digital yang Bermanfaat Bagi Bali Dimasa Pandemi. In *SANDI: Seminar Nasional Desain* (Vol. 1, pp. 158–165).

Nwagbara, U., & Reid, P. (2013). Corporate social responsibility communication in the age of new media: Towards the logic of sustainability communication. *Revista de Management Comparat International, 14*(3), 400.

Satya, M. T., Tejaningrum, A., & Hanifah, H. (2021). Wisata virtual kampung kreatif cibunut finest. *BERNAS: Jurnal Pengabdian Kepada Masyarakat, 2*(4), 926–935.

Susanto, E. B., Wijonarko, R., Purwanto, E., & Binabar, S. W. (2019). Kampung batik digital berbasis virtual tour sebagai wadah promosi batik Kota Pekalongan di era industri 4.0 (studi kasus: Kampung Batik Banyurip). *Jurnal Litbang Kota Pekalongan, 17*, 69–70.

Validation Preparation Final Report (2019). UNESCO.

Yan, Z., Ma, X., & Tao, S. (2009). The virtual display study on 3D panorama in tourist areas-take Shilin world geopark as an example. In Proceedings. The 2009 International Symposium on Information Processing (ISIP 2009) (p. 229). Academy Publisher.

9

GAMIFICATION AND SUSTAINABLE BEHAVIOUR

A Case Study of China

Huang Miao, Mohamad Saifudin Mohamad Saleh and Izzal Asnira Zolkepli

Introduction: Sustainability Issues in China

Rapid economic development has quickly driven China to become the second largest economy of the world over recent decades (Zhang et al., 2017). However, severe environmental problems, especially air pollution from carbon dioxide (CO_2), sulphur dioxide (SO_2), and soot emissions, gradually become prominent in this process. Although the government has made tremendous regulatory efforts, such problems still threaten national development and public health (Zhang et al., 2017). For instance, industrial growth has brought serious pollution to the three major ecosystems – soil, water, and air – while the ever-increasing population has also brought tremendous pressure to bear on the Earth's environment (Khan & Chang, 2018). Over recent years, the widespread consumption of fossil fuels has emitted huge amounts of greenhouse gases and delivered severe climate change to China.

Committed to achieving the sustainable development goal of carbon neutrality by 2060, China regards reducing carbon emissions as an important mission (Xu et al., 2021). In addition, China has also vigorously boosted afforestation projects to expand forest coverage, in order to curb air pollution and climate change (Chen et al., 2019). Since 1978, to combat desertification, the government has initiated many large-scale environmental protection projects to promote vegetation restoration, desert management, water source management, and comprehensive agricultural development (Bryan et al., 2018), most of which are very effective. However, facing the complexity of environmental issues, the Chinese government often adopts a top–down performance supervision strategy, without sufficient research on bottom–up public supervision methods (Wu et al., 2018).

Coupled with highly intensive environmental pressure, China's urbanisation process, rural poverty alleviation, and economic transformation are also

DOI: 10.4324/9781003309642-9

encountering huge sustainability challenges. Since urban land was allowed to be commercially developed in 1987, the income from leasing and trading of local urban land in China has skyrocketed, because local governments have intrinsic motivations to expand land leases and transactions, which are of great significance for expanding local fiscal revenue (Chen et al., 2018). However, the disorderly urbanisation from large-scale conversion of rural land to urban land has given birth to many new crises: farmland destruction, environmental degradation, food insecurity, and so on (Chen et al., 2018). China's rapid urbanisation has caused serious land degradation and population poverty in rural areas, thus disrupting the balanced development of society and economy (Liu & Wang, 2019). Therefore, it is necessary to curb urban expansion; build an urban ecosystem with reasonable structures, complete functions, and efficient processes; and ensure high standards of residential areas. These measures are important breakthroughs in sustainable urbanisation strategies (Zhang et al., 2020). It is critical to improve land structure and quality, innovate agricultural technologies, and promote the coupling of people and land systems. These approaches are necessary means to alleviate the poverty of the rural population and ensure their quality of life (Liu & Wang, 2019; Guo & Liu, 2022).

In addition, in order to boost investment in clean energy, local governments will encourage the employment of green finance strategies in the financial sector in China. With green manufacturing as the core, local governments will follow rational industrialisation policies that take into account economic growth, environmental protection, and productivity (Zahoor et al., 2021). As a long-term, positive development model, sustainable economic development entails not only a suitable speed of economic growth but also high-quality economic development, as well as harmonies between the economy, nature, and society. For China, with a large population, uneven resources, and great pressure on employment and environmental protection, sustainable economic growth means the transformation of the growth mode, involving the optimisation of industrial structure, the improvement of scientific and technological innovation capabilities, the enhancement of labour quality, and the balance of regional economy (Guo & Ke, 2020).

Requirement for Sustainable Mobile Applications in China

With the popularity of the mobile internet and mobile applications, Chinese mobile application developers are trying to meet the increasingly diversified needs of internet users. In particular, the implementation of mobile applications to manage environmentally sustainable projects has gradually drawn wide attention in the field of academic research. These needs are mainly showcased in the following areas: (1) recycling management, (2) environmental monitoring, (3) environmental education, and (4) energy management.

In response to the above demands, researchers have put forward some successful cases. The inefficiency of municipal solid waste (MSW) collection systems in large cities makes it hard for individual users to understand MSW's recycling channels. As a result, a new "internet + recycling" waste recycling model has been proposed on an online platform, which allows collectors to make collection appointments through mobile applications and then make on-site collections (Gu et al., 2019).

In addition, in order to formulate countermeasures, environmental management requires multipoint monitoring of environmental information. To meet these needs, a mobile application integrates Raspberry Pi and Arduino chips to perform four-level system monitoring for agricultural greenhouse environments (Geng et al., 2019). With excellent motion control and mobile acquisition functions, this application can capture crop pictures at low cost and then upload them to a data server for analysis, hence, realising multipoint monitoring of agricultural environments (Geng et al., 2019).

In the education sector, there is a strong demand for mobile augmented reality (AR) and gamification to facilitate environmental education. For example, a scavenger hunt game has been designed based on geographic location information and mobile AR technology, intended to explore the impact of embedded digital technology and gamification on environmental education in Chinese higher education in languages (Mei & Yang, 2019). The successful application of the game shows that combining mobile technology with gamification can successfully boost environmental education (Mei & Yang, 2019).

Furthermore, strengthening energy management and ensuring energy security are common social demands for sustainable mobile applications. The Chinese government provides reliable, cheap, and sustainable electricity services, and precise energy management is a foundation for securing these services. Therefore, for the security of power grid systems, geographic information from mobile applications may help establish detection and prevention strategies accordingly (Zhou et al., 2017).

Role of Gamification in Promoting Sustainable Behaviour in China

Deterding et al. (2011) defined gamification as "the use of game design elements in non-game contexts". Hamari et al. (2014) extended this definition to motivational affordances, psychological outcomes, and behavioural results. Many companies are using gamification as a tool to stimulate consumer behaviour. In fact, many gamification elements, such as avatars, badges, points, levels, virtual gifts, and so on, are very easy to identify and can influence consumers so as to encourage them to enter certain states of mind (Gatautis et al., 2016). At the same time, meanings, social influence, ownership, credibility, achievability, and challenges are also regarded as the core elements of gamification applications to strengthen the physical and mental connections between users and applications (Ouariachi & Elving, 2020).

On the other hand, China's green and low-carbon behaviours have received extensive attention from academia, such as household energy-saving practices

(Zhao et al., 2019), low-carbon consumption (Mi et al., 2019), and employee energy-saving plans (Tang et al., 2019). In the theory of planned behaviour, researchers have established a set of behavioural models to analyse the willingness of Chinese consumers to purchase green appliances (Wang et al., 2017). Also, by using this theory, Liang et al. (2019) studied Chinese consumers' purchasing behaviour for green products from the emotional perspective. In addition, public information is also found to have played an important role in China's green-housing purchasing behaviour (Zhang et al., 2016).

At present, to deliver environmental education with the most extensive influence, Chinese developers have added gamification elements to non-game scenarios to guide users' sustainable behaviour, because adding such elements (e.g., feedback, challenges, and points) to traditional instructional settings can hike students' engagement and motivation and help them achieve certain learning outcomes (Khan et al., 2017). For example, by developing a "discovery map" for a botanical garden in Southwestern China and adding games to visitors' tours, researchers have significantly enhanced their understanding of ecosystem diversity (Yang & Chen, 2017).

Based on the goal frame theory, Lin and Du (2021) constructed a user participation model and proposed a formation and cultivation mechanism for sustainable behaviour in a green gamification platform, finding that high-intelligence individuals do not need excessive participation in the process of forming and cultivating sustainable behaviours. Their research also confirmed that the application of gamification design can deliver on the goal of attracting and guiding users' sustainable behaviour (Lin & Du, 2021). In terms of gamification applications, the game modules of Alipay and *Ant Forest* are leading China's mobile applications.

Alipay as a Fintech Company with Sustainable Elements in Mobile Applications

Technological innovation and integration with finance have driven the development of financial technology (fintech) since the 2008 global financial crisis (Shim & Shin, 2016). As the most cutting-edge and deepest technological innovation in the financial field (Schindler, 2017), fintech covers blockchain, cryptocurrency, digital trading, deep learning and artificial intelligence, peer-to-peer lending, mobile payment, and equity crowdfunding (Philippon, 2016). Today, more and more fintech companies are engaged in various ways with environmental projects that are making increasingly greater contributions to sustainable development (Folke et al., 2019; Galaz et al., 2018).

In March 2015, the Chinese government clearly set forth the initiative of "internet +" in its government work report, indicating that China had begun to promote the application of the internet and other information technologies in traditional industries at the government level. Such implementation mainly includes mobile internet, big data, cloud services, and the Internet of Things, which can all help to transform and upgrade traditional industries (State Council,

2015). For example, Alibaba and Tencent, two giant companies with a huge market value in China, deliver a significant contribution to achieving sustainable development, relying on their massive user bases and advanced internet and big-data technologies. In fact, Alibaba and its subsidiaries, Alipay and *Ant Forest*, have begun to demonstrate the potentials of their business models in environmental protection and land resource restoration (Xiong & Meng, 2018).

In April 2019, Alipay announced that *Ant Forest* had appealed to 500 million users and planted more than 100 million trees in desert regions, covering a total area of 100,000 hectares. In June 2019, Alipay launched the Public Welfare Forest project to attract university students and volunteers to plant trees for their organisations. This move enabled more enterprises and institutions in China to participate in environmental protection projects and extended this cause to wider audiences. In August 2019, Alipay released a research report on public low-carbon lifestyles in the context of the internet, which claimed that, by the third anniversary of the founding of *Ant Forest*, its 500 million users had planted 122 million real trees, covering an area equivalent to 1.5 times the size of Singapore and cutting carbon emissions by 7.92 million tons (Zhang et al., 2021).

On the basis of the success of *Ant Forest*, Alipay has launched more game functions in *Ant Manor*. As in *Ant Forest*, users of Alipay can receive in-game farm fodder via mobile payments, which can then be used to feed virtual pet chicks. As the chicks grow up, users get virtual reward "eggs", which can be converted into materials for donation to children in impoverished areas (Figure 9.1). The success

FIGURE 9.1 Interface of *Ant Manor*.

of *Ant Forest* and *Ant Manor* highlights fintech's achievements in terms of scientific and technological power: while following business rules, they can advance public welfare and promote sustainable economic and social development. Under the leadership of the Alipay model, many more fintech companies are now committed to the development of green gamification applications – for example, the Panda Protection Project, launched by Sina Weibo in 2018, which is called *Forest Station*, and *Forest*, a mobile application aimed to improve concentration by urging users to put down their mobile phones and focus on things at hand, based on the concept of planting trees (Qu, 2019).

Roles of *Ant Forest*'s Mobile Application in Boosting Environmental Sustainability

In 2016, Alipay launched the game module – *Ant Forest* – which allows individuals to establish carbon sub-accounts under their general accounts and earn "green points" by collecting and sharing low-carbon behaviours (Yang et al., 2018). Green points can be exchanged for 24 kinds of daily sustainable behaviours, such as mobile payment, green travel, and waste recycling. At the same time, the game's social functions enable users to give or get points or choose to cooperate with others in planting trees (Wang & Yao, 2020). When the points reach a certain target, users can plant a real sapling in the desert through the company's agency (Wang & Yao, 2020). This information-based game module encourages users to record their own carbon footprint and reduce their personal carbon emissions, thus gradually promoting a low-carbon lifestyle (Zhang et al., 2020). For this reason, *Ant Forest* won the highest United Nations Environmental Action award in 2019 – the Champion of the Earth Award (Rukikaire, 2019).

As an environmentally friendly gamification application, *Ant Forest* supports its users to enjoy the fun of the game and get a sense of satisfaction by earning green points in the leader-board system and planting trees both in the virtual and real worlds (Zhang et al., 2020). Other studies have further discovered that *Ant Forest* can meet users' expectations and alleviate their anxiety about the environment (Zhang et al., 2020). Social interaction is actually an important gaming experience (Arango-López et al., 2021). In *Ant Forest*, users can visit their friends' spaces through social interaction functions to obtain or help collect others' green energy points (Zhang et al., 2020). However, energy must be harvested every day, and it needs to be operated one at a time. Growing a sapling requires users to invest considerable time and effort, which may dampen their interest and make them feel impatient, finally reducing their satisfaction (Zhang et al., 2020).

In summary, as a representative of environmentally friendly mobile applications, *Ant Forest* has an impact on users' sustainable behaviour in two main aspects: advocating a low-carbon lifestyle and encouraging afforestation (Li & Peng, 2019). In this way, *Ant Forest* encourages users to participate in daily sustainable behaviours and social interactions so as to illustrate them as green benefits. Users record their carbon footprints in the application, collect green energy points, exchange them for tree-planting opportunities in desert areas, and solidify their willingness to fulfil their sustainability obligations. This shows that *Ant Forest* is able to integrate persuasive systems, social media, and gamification entertainment to elicit sustainable behaviour (Yang et al., 2018).

Meaningful and Reward-based Gamification Elements in *Ant Forest*

Reiners and Wood (2015) distinguish two types of gamification: meaningful gamification and reward-based gamification. The former is regarded as an internal motivation that can change learning models, promote learners to fuse their previously held beliefs, and drive these beliefs' long-term transformation (Liu et al., 2017). In contrast, the latter is a system that adjusts behaviours by adding badges, levels, leader boards, achievements, points, and so on, in a virtual environment (Reiners & Wood, 2015); it is often regarded as an extrinsic motivation, with a short-term effect on users' behaviours. In other words, once the motivation stops, the behaviour will also be terminated (Skinner, 1938).

Oliver et al. (2015) clarified the concept of meanings of gamification elements: the meaning is not a simple positive or negative evaluation, but a mixed and unique emotion – for example, being moved or touched. Such emotions can be integrated into game functions in such forms as in-depth stories, interactions between characters, challenges and competitions, and communications between players (Rogers et al., 2017). The creation of meanings comes from balancing the psychological needs of injustice in the world while paying attention to the relationship between human virtues, emotions, and values. Non-material rewards and compensation for the characters in stories can often satisfy people's cognitive needs for meaning (Raney et al., 2019).

Based on the concept of play and participatory museums, Nicholson (2015) put forward the following six meaningful gamification strategies: play, choice, exposition, information, engagement, and reflection. "Play" facilitates users to create and explore activities freely. Then, "choice" provides users with a selection of ways, times, and places to participate in games. "Exposition" provides a narrative function for meaningful gamification, so as to maintain game players' strong interest in game stories and increase their curiosity and excitement for games. It can also help players find the path of clearance, explain operations and techniques, and eliminate ambiguities about goals. "Information" forms a link between users and systems, and it can be used to transfer knowledge on operations and systems more effectively. "Engagement" refers to social participation by players, which can be initiated through such channels as chatting, forums, and sharing, so as to maintain healthy competition and cooperative relationships. Finally, "reflection" allows players to re-examine their game history and form more thoughts (Nicholson, 2015).

In the gameplay of *Ant Forest*, users can choose whether to participate in a project to trigger the "play" experience; afterwards, they can make a "choice" of game difficulty by selecting the trees they want to plant. In the process of the game, *Ant Forest* will inform users of the growth of their trees, the real tree-planting environment, and environmental protection stories, so as to provide enough information and background stories to mobilise the emotions of players, who can find the interests and meanings in sustainable behaviours and practise them. At the same time, *Ant Forest* subtly teaches users sustainability knowledge and guides them to think more about environmental protection (Figure 9.2). During the entire game process, users can actively share their game progress and achievements through social networks, thus greatly improving the "engagement" between group users and laying a basis for attracting more users for the operation of *Ant Forest.*

In addition, Phillips et al. (2013) proposed the following definition for rewards in gamification elements: they are a positive return that serves to reinforce players' behaviour within a game. The rewards here specifically refer to things that can be measured in games, excluding the activities that are considered "rewards" but without return in daily life (for example, the fun of playing games; Phillips et al., 2013). Although the role of rewards as a gamification element in the educational environment is still controversial (Seaborn & Fels, 2015), several benefits of rewards have been identified in the educational learning environment, including internal and external motivation (Nicholson, 2015), a sense of progress, achievement of goals (Hamari, 2017), or just fun (Codish & Ravid, 2014). Rewards in the online environment include a variety of gamification elements,

FIGURE 9.2 Interfaces for selecting trees and information about protected areas in *Ant Forest.*

such as points, badges, or levels (Ortega-Arranz et al., 2017), which are designed to hike the enthusiasm of participants.

A study by Phillips et al. (2015) invited a focus group composed of game designers, researchers, and players to investigate the reward types in more than 60 commercial video games, in which they observed a total of 915 reward instances. These instances were then classified into five reward types, namely access, facility, sustainability, glory, and praise communication. Besides, there is an additional reward modifier, called sensory feedback. Rewards of "access" refer to helping players enter other inaccessible environments, objects, or modes (including cutscenes). Rewards of "facility" refer to increasing players' efficiency in a game. Rewards of "sustainability" refer to reducing the burden on players and mitigating negative game states. Rewards of "glory" are considered to be rewards that do not affect real-time gameplay but are quantified after the game. Rewards of "praise communication" refer to the form of praise or flattery for players through the game system. Although sensory feedback is not a reward in itself, it provides players with tactile feedback or aesthetics that can directly modify the rewards (Phillips et al., 2015).

Ant Forest also offers diverse rewarding elements. By just performing small, simple operations every day, users can complete their long-term tasks, with each effort instantly illustrated in the growth of virtual trees or commodity discounts in games in an intuitive form. After a period of gameplay, these rewards eventually convert into real trees to give users additional spiritual rewards (Figure 9.3).

FIGURE 9.3 Interfaces for watering trees and getting badges in *Ant Forest*.

With a system of points and badges, *Ant Forest* also ensures that users own the capital to show off or compare with each other. These rewards can also be synchronised to other social networks, allowing users to receive more honourable rewards, such as praise and flattery.

Conclusion

The rapid development of the economy has gradually led to serious environmental problems in China, including carbon emissions caused by fossil fuels, desertification caused by water shortages, and smog in urban areas; these have become environmental crises and affect public health and social development. Coupled with these environmental pressures, China's urbanisation process, rural poverty alleviation, and economic transformation are also encountering huge sustainability challenges. To address these problems, the Chinese government has put forward a sustainable development goal of achieving carbon neutrality by 2060, with some specific measures being taken vigorously in recent years, such as afforestation, desert management, water resource management, and comprehensive agricultural development.

In the fields of mobile internet and mobile applications, the application requirements for sustainable development have been proposed and are mainly concentrated in four areas: recycling management, environmental monitoring, environmental education, and energy management. To better meet these requirements, mobile app developers are trying to incorporate gamification into their products so as to promote sustainable behaviour of the public in China.

As fintech appeals to a large number of users, high-tech internet companies make significant contributions to China's sustainable development. In particular, gamified mobile applications represented by Alipay and its game module, *Ant Forest*, have educated more than 500 million users, with more than 100 million real trees planted in desert areas across China.

Taking *Ant Forest* as a successful case, this chapter mainly explores the promotion of meaningful engagements and diverse rewards, which are regarded as gamification elements in its gameplay. Tightly integrated with the functionality of the mobile application, such elements have fostered sustainable behaviours among massive numbers of users through their daily mobile usage.

Acknowledgement

This work was funded by Jiangsu University Philosophy and Social Science Research Project, under the Jiangsu Provincial Department of Education and Communication University of China, Nanjing. Project title: The Impact of Interactive Applications on Users' Carbon Neutral Behavior in the Context of Mobile Network (2021SJA2319).

References

Arango-López, J., Vela, F. L. G., Rodríguez, P. P., Moreira, F., & Fonseca, D. (2021). Using geolympus to create a pervasive game experience in the higher education context. *Universal Access in the Information Society, 20*(1), 1–13.

Bryan, B. A., Gao, L., Ye, Y., Sun, X., Connor, J. D., Crossman, N. D., Stafford-Smith, M., Wu, J., He, C., Yu, D., Liu, Z., Li, A., Huang, Q., Ren, H., Deng, X., Zheng, H., Niu, J., Han, G., & Hou, X. (2018). China's response to a national land-system sustainability emergency. *Nature, 559*(7713), 193–204.

Chen, C., Park, T., Wang, X., Piao, S., Xu, B., Chaturvedi, R. K., Fuchs, R., Brovkin, V., Ciais, P., Fensholt, R., Tømmervik, H., Bala, G., Zhu, Z., Nemani, R. R., & Myneni, R. B. (2019). China and India lead in greening of the world through land-use management. *Nature Sustainability, 2*(2), 122–129.

Chen, J., Guo, F., Wang, H., Wang, Z., & Wu, Y. (2018). Urban land revenue and sustainable urbanization in China: Issues and challenges. *Sustainability, 10*(7), 2111.

Codish, D., & Ravid, G. (2014). Academic course gamification: The art of perceived playfulness. *Interdisciplinary Journal of E-Learning and Learning Objects, 10*(1), 131–151.

Deterding, S., Dixon, D., Khaled, R., & Nacke, L. (2011). From game design elements to gamefulness: defining "gamification". In Proceedings of the 15th international academic MindTrek conference: Envisioning future media environments, 9-15.

Folke, C., Österblom, H., Jouffray, J. B., Lambin, E. F., Adger, W. N., Scheffer, M., Crona, B. I., Nyström, M., Levin, S. A., Carpenter, S. R., Anderies, J. M., Chapin, S. III., Crépin, A. S., Dauriach, A., Galaz, V., Gordon, L. J., Kautsky, N., Walker, B. H., Watson, J. R., Wilen, J., & de Zeeuw, A. (2019). Transnational corporations and the challenge of biosphere stewardship. *Nature Ecology & Evolution, 3*(10), 1396–1403.

Galaz, V., Crona, B., Dauriach, A., Scholtens, B., & Steffen, W. (2018). Finance and the Earth system–Exploring the links between financial actors and non-linear changes in the climate system. *Global Environmental Change, 53*, 296–302.

Gatautis, R., Vitkauskaite, E., Gadeikiene, A., & Piligrimiene, Z. (2016). Gamification as a mean of driving online consumer behaviour: SOR model perspective. *Engineering Economics, 27*(1), 90–97.

Geng, X., Zhang, Q., Wei, Q., Zhang, T., Cai, Y., Liang, Y., & Sun, X. (2019). A mobile greenhouse environment monitoring system based on the internet of things. *IEEE Access, 7*, 135832–135844.

Gu, F., Zhang, W., Guo, J., & Hall, P. (2019). Exploring "Internet+ Recycling": Mass balance and life cycle assessment of a waste management system associated with a mobile application. *Science of the Total Environment, 649*, 172–185.

Guo, B., & Ke, J. (2020). The impacts of high-speed rail on sustainable economic development: evidence from the central part of China. *Sustainability, 12*(6), 2410.

Guo, Y., & Liu, Y. (2022). Sustainable poverty alleviation and green development in China's underdeveloped areas. *Journal of Geographical Sciences, 32*(1), 23–43.

Hamari, J. (2017). Do badges increase user activity? A field experiment on the effects of gamification. *Computers in Human Behavior, 71*, 469–478.

Hamari, J., Koivisto, J., & Sarsa, H. (2014). Does gamification work? - a literature review of empirical studies on gamification. In 2014 47th Hawaii international conference on system sciences, 3025-3034.

Khan, A., Ahmad, F. H., & Malik, M. M. (2017). Use of digital game based learning and gamification in secondary school science: The effect on student engagement, learning and gender difference. *Education and Information Technologies, 22*(6), 2767–2804.

Khan, M. I., & Chang, Y. C. (2018). Environmental challenges and current practices in China – a thorough analysis. *Sustainability, 10*(7), 2547.

Li, L., & Peng, Z. (2019). Research on sustainable development – take "ant forest" for example. *Earth and Environmental Science, 242*, 052031.

Liang, D., Hou, C., Jo, M. S., & Sarigöllü, E. (2019). Pollution avoidance and green purchase: The role of moral emotions. *Journal of Cleaner Production, 210*, 1301–1310.

Lin, Y. X., & Du, H. S. (2021). An analysis on the formation and cultivation of environmental protection norms in the context of green gamification. *Proceedings of The International Conference on Electronic Business, 21*, 53–64. Accessed from https://aisel.aisnet.org/iceb2021/38

Liu, D., Santhanam, R., & Webster, J. (2017). Toward meaningful engagement: a framework for design and research of gamified information systems. *MIS Quarterly, 41*, 1011–1034.

Liu, Y., & Wang, Y. (2019). Rural land engineering and poverty alleviation: Lessons from typical regions in China. *Journal of Geographical Sciences, 29*(5), 643–657.

Mei, B., & Yang, S. (2019). Nurturing environmental education at the tertiary education level in China: can mobile augmented reality and gamification help? *Sustainability, 11*(16), 4292.

Mi, L., Zhu, H., Yang, J., Gan, X., Xu, T., Qiao, L., & Liu, Q. (2019). A new perspective to promote low-carbon consumption: The influence of reference groups. *Ecological Economics, 161*, 100–108.

Nicholson, S. (2015). A recipe for meaningful gamification. *Gamification in Education and Business*, 1–20. Cham Heidelberg New York Dordrecht London: Springer.

Oliver, M. B., Bowman, N. D., Woolley, J. K., Rogers, R., Sherrick, B. I., & Chung, M. Y. (2015). Video games as meaningful entertainment experiences. *Psychology of Popular Media Culture, 5*(4), 390–405.

Ortega-Arranz, A., Munoz-Cristóbal, J. A., Martínez-Monés, A., Bote-Lorenzo, M. L., & Asensio-Pérez, J. I. (2017). How gamification is being implemented in MOOCs? A systematic literature review. In European Conference on Technology Enhanced Learning, 441–447. Accessed from http://uvadoc.uva.es/handle/10324/24837

Ouariachi, T., & Elving, W. (2020). Accelerating the Energy Transition through Serious Gaming: Testing Effects on Awareness, Knowledge and Efficacy Beliefs. *Electronic Journal of e-Learning, 18*(5), 410–420.

Philippon, T. (2016). *The fintech opportunity* (p. 22476). National Bureau of Economic Research. Accessed from https://www.nber.org/papers/w22476

Phillips, C., Johnson, D., & Wyeth, P. (2013). Videogame reward types. In Proceedings of the First International Conference on Gameful Design, Research, and Applications, 103–106.

Phillips, C., Johnson, D., Wyeth, P., Hides, L., & Klarkowski, M. (2015). Redefining videogame reward types. In Proceedings of the Annual Meeting of the Australian Special Interest Group for Computer Human Interaction, 83–91.

Qu, S. (2019). The impact of college students' participation in online environmental protection games on environmental protection behavior: the mediating role of environmental values and environmental self-responsibility. Shaanxi Normal University. Accessed from https://cdmd.cnki.com.cn/Article/CDMD-10718-1019223631.htm

Raney, A. A., Oliver, M. B., & Bartsch, A. (2019). Eudaimonia as media effect. In *Media effects: Advances in theory and research* (4th edition, pp. 258–274). New York: Routledge.

Reiners, T., & Wood, L. C. (2015). *Gamification in Education and Business.* Cham Heidelberg New York Dordrecht London: Springer.

Rogers, R., Woolley, J., Sherrick, B., Bowman, N. D., & Oliver, M. B. (2017). Fun versus meaningful video game experiences: A qualitative analysis of user responses. *The Computer Games Journal, 6*(1–2), 63–79.

Rukikaire, K. (2019). *Chinese initiative ant Forest wins UN champions of the earth award.* UNEP – UN Environ. Programme. Accessed from http://www.unenvironment.org/news-and-stories/press-release/chinese-initiative-ant-forest-wins-un-champions-earth-award

Schindler, J. W. (2017). FinTech and financial innovation: Drivers and depth. *Finance and Economics Discussion Series, 2017-81,* 1–16.

Seaborn, K., & Fels, D. I. (2015). Gamification in theory and action: A survey. *International Journal of Human-Computer Studies, 74,* 14–31.

Shim, Y., & Shin, D. H. (2016). Analyzing China's fintech industry from the perspective of actor–network theory. *Telecommunications Policy, 40*(2–3), 168–181.

Skinner, B. F. (1938). *The behavior of organisms: An experimental analysis.* England: Appleton.

State Council. (2015). *Guiding opinions of the state council on actively promoting internet+ action.* Central People's Government of the People's Republic of China. Accessed from http://www.gov.cn/zhengce/content/2015-07/04/content_10002.htm

Tang, Z., Warkentin, M., & Wu, L. (2019). Understanding employees' energy saving behavior from the perspective of stimulus-organism-responses. *Resources, Conservation and Recycling, 140,* 216–223.

Wang, X., & Yao, X. (2020). Fueling pro-environmental behaviors with gamification design: identifying key elements in ant forest with the kano model. *Sustainability, 12*(6), 2213.

Wang, Z., Wang, X., & Guo, D. (2017). Policy implications of the purchasing intentions towards energy-efficient appliances among China's urban residents: Do subsidies work?. *Energy Policy, 102,* 430–439.

Wu, J., Xu, M., & Zhang, P. (2018). The impacts of governmental performance assessment policy and citizen participation on improving environmental performance across Chinese provinces. *Journal of Cleaner Production, 184,* 227–238.

Xiong, J., & Meng, Q. (2018). The analysis of ant forest business model: Under "Internet Green Finance". *Open Access Library Journal, 5,* e4887.

Xu, C., Yang, J., He, L., Wei, W., Yang, Y., Yin, X., ... & Lin, A. (2021). Carbon capture and storage as a strategic reserve against China's CO_2 emissions. *Environmental Development, 37,* 100608.

Yang, X., & Chen, J. (2017). Using discovery maps as a free-choice learning process can enhance the effectiveness of environmental education in a botanical garden. *Environmental Education Research, 23*(5), 656–674.

Yang, Z., Kong, X., Sun, J., & Zhang, Y. (2018). Switching to green lifestyles: Behavior change of ant forest users. *International Journal of Environmental Research and Public Health, 15*(9), 1819.

Zahoor, Z., Khan, I., & Hou, F. (2021). Clean energy investment and financial development as determinants of environment and sustainable economic growth: Evidence from China. *Environmental Science and Pollution Research, 29,* 16006–16016.

Zhang, D., Xu, J., Zhang, Y., Wang, J., He, S., & Zhou, X. (2020). Study on sustainable urbanization literature based on Web of Science, scopus, and China national knowledge infrastructure: A scientometric analysis in CiteSpace. *Journal of Cleaner Production, 264*, 121537.

Zhang, J., Cheng, M., Mei, R., & Wang, F. (2020). Internet use and individuals' environmental quality evaluation: evidence from China. *Science of The Total Environment, 710*, 136290.

Zhang, L., Sun, C., Liu, H., & Zheng, S. (2016). The role of public information in increasing homebuyers' willingness-to-pay for green housing: Evidence from Beijing. *Ecological Economics, 129*, 40–49.

Zhang, M., Sun, X., & Wang, W. (2020). Study on the effect of environmental regulations and industrial structure on haze pollution in China from the dual perspective of independence and linkage. *Journal of Cleaner Production, 256*, 120748.

Zhang, Q., Zhang, S., Ding, Z., & Hao, Y. (2017). Does government expenditure affect environmental quality? Empirical evidence using Chinese city-level data. *Journal of Cleaner Production, 161*, 143–152.

Zhang, Y., Chen, J., Han, Y., Qian, M., Guo, X., Chen, R., Xu, D., & Chen, Y. (2021). The contribution of Fintech to sustainable development in the digital age: Ant forest and land restoration in China. *Land Use Policy, 103*, 105306.

Zhao, X., Cheng, H., Zhao, H., Jiang, L., & Xue, B. (2019). Survey on the households' energy-saving behaviors and influencing factors in the rural loess hilly region of China. *Journal of Cleaner Production, 230*, 547–556.

Zhou, C., Feng, R., Wang, L., Huang, W., & Guo, Y. (2017). Security attack detection algorithm for electric power gis system based on mobile application. *IOP Conference Series: Earth and Environmental Science, 64*, 012101.

10

FACTORS ENCOURAGING JOURNALISTS IN PURSUING SUSTAINABILITY COMMUNICATION

Anna Agustina and Adnan Hussein

Introduction: Indonesian Media and Journalists' Landscape

Indonesia has become the host of a large volume of media on various platforms. Previous research stated that, in Indonesia, there are over 1200 printed publications, more than 900 radio stations, hundreds of national and regional television channels (Lim, 2011, 2012), and around 43,400 forms of online media (Ambardi et al., 2014). In addition, there are social media and citizen journalism that enlarge the volume of news and messages distributed on events on a daily basis. Along with the number of media, there are also thousands of journalists, and millions of people who produce news and create media content (Dhyaksa, 2017; Tesalonika, 2020). Both news and more general media content can be accessed by the people not only via a synchronous mode but also via an asynchronous one. Both the coverage and the public who view the content are local, national, and international.

One important issue in the Indonesian media landscape is the ownership of hundreds of mainstream media and thousands of online media. Merlyna Lim (2011) published a report titled 'The League of Thirteen' that described media concentration in Indonesia. The media in Indonesia are generally owned by 13 businessmen, and almost all prioritise the interest of the owner, which is in business schemes managing media organisations, rather than promoting journalistic idealism (Lim, 2012). This arrangement tends to situate the process of journalism as a business orientation, instead of it being executed with journalism values. In essence, the media are more concerned with business matters than journalistic values. This situation is challenging to journalists in their pursuit of sustainability communication that should balance economic, social, and environmental factors. Added to that, from the media management point of view, journalists are requested to multitask as generalists rather than specialists, in the

DOI: 10.4324/9781003309642-10

transformation into a more effective and efficient workforce in the digital era. The structure of media organisations tends to be more efficient, with the merging of several mainstream media into one newsroom in Indonesia since the 1990s.

Journalists are categorised in several ways. Daeng, a senior journalist in Kalimantan, states that not only are journalists clustered according to common issues, they are also assembled based on coverage area, such as regional versus central journalists, and the roles they play, such as government offices journalists for journalists who are based in the government pressroom, or celebrity and entertainment journalists for journalists who are based in production houses or film agencies and always report on entertainment, lifestyle, or celebrity activities (personal communication, 18 May 2021).

Noni, an environmental journalist in Java, explains that there are also permanent journalists and freelance journalists. The permanent journalists are those who work only for one media organisation. Freelance journalists, on the other hand, write news based on assignments from several media organisations. While fulfilling the role of a journalist, freelancers could also work as a member of an NGO. Other than that, they are able to engage in environmental journalism based on their passion and research, following procedural reporting, and then submit their work to several media organisations. Any interested organisation can then confirm its interest and acceptance for publication on a first come, first served basis.

In the various schemes described above, it is the journalists who decide the kind of event they wish to cover and the media they want to do it for. It is also the journalists who provide the perspectives on an event or issue and report them accordingly. Some journalists are given the liberty and authority to execute the whole journalism process from reporting, writing, and editing to publishing the final news directly on the platform. This type of journalistic work and the important role that journalists play in covering events were also described by Donsbach in 2004 and are still relevant at the present time. Journalists are, therefore, the focus of this chapter. They are the main actors in news production and act as the content creators of media messages. This chapter explores the factors that influence journalists to execute environmental communication through journalism within the context of sustainability communication in Indonesia. It further describes the kind of content a journalist posts in the media.

The Importance of Ecological Content

In the year 2020, an American journalist called Philip Jacobson was deported from Indonesia on either one of two grounds. One report stated that Jacobson wrote about sensitive aspects of Indonesia's environmental problems, while another asserted that he bent an administrative rule by not having a journalist visa while acting as a journalist. The same also happened to environmental journalists from France in 2014 and the UK in 2016. They were deported for filming a documentary in West Papua and working as journalists without a journalist visa. In respect to the rules and regulations that should be obeyed, three journalists from different countries reported environmental-related issues that are sensitive. MacAndrews (1994) research results described that environmental issues in Indonesia become sensitive because it is related to the interests of the government in power and business entities.

The issues are deemed sensitive as they relate to the interests of many stakeholders. Reporting on them means upsetting the interests of many stakeholders, including the government in power, and their network. On the other hand, environmental problems in Indonesia are getting worse. We can see the shrinking land of the Indonesian archipelago in the middle of the city. Jalan Sudirman, the main business district in the capital city of Jakarta, has a pedestrian bridge that crosses a road. At present, there is a space of approximately 5 cm between the bridge and the street, indicating structural environmental issues. Indonesia has also experienced acid rain since 2019 that has impacted the quality and quantity of crops, farming, and fisheries harvests. Therefore, within a challenging context, journalists must still conduct sustainability communication by carrying out environmental journalism and posting environmental news to raise awareness and help find solutions to these problems.

Notwithstanding, Indonesian journalists are known to actively play important roles in contributing solutions to national social problems whichever government is in power. They always find ways to present journalists' idealism to the public. Therefore, as well as reporting and producing news content that supports sustainability within the current demanding context for the government, they must also do so for the business owners who manage media organisations in the business world.

Sustainability Communication in Indonesian Journalism

The environment is a collectively shared resource, and natural resources are finite and limited, whereas human interactions and needs are infinite and keep on growing. Within the context, it is necessary that we propose strategies for our usage of natural resources to sustain nature and non-renewable properties. At the same time, the natural environment is a complex system. While it can be controlled by humans, there are so many interests involved in the use of natural resources that turn manipulative and exploitative. One of the strategies to eradicate this involves discussing and spreading the word about conserving natural resources, adopting environmentally friendly behaviours, or mitigating the problems based on the role of each stakeholder in society. Sustainability communication through the work of journalism is one of the efforts to realise this.

Sustainability communication draws on a wide range of disciplines that have the body of knowledge and methodology approaches to illuminate the drivers of and barriers to a broader and deeper societal engagement with the idea of sustainability (Fischer et al., 2016, p. 3). In line with this, Mohamad Saifudin (2017) also highlights that sustainability communication has the goal of enhancing society's awareness of environmental issues (p. 92). Therefore, it is the authors' aim to describe the factors that are important for journalists pursuing sustainability communication by interviewing journalists who claim to have already been producing environmental communication for at least 10 years.

The process of identifying the factors that are important for journalists contributing to sustainability communication was originally conducted in 2014, involving 30 environmental journalists. Because of the pandemic, the data update conducted in 2021 could only involve 15 environmental journalists. All 15 informants are members of the Society of Indonesian Environmental Journalists (SIEJ). They are based in Jakarta (five), Yogya (one), Semarang (one), Aceh (two), Balikpapan/Kalimantan (three), Makassar (two), and Soppeng (one). The data collection techniques were in-depth interviews via long-distance video calls to journalists outside Jakarta and face-to-face interviews with journalists residing in Jakarta. This chapter focuses on the journalists' perspectives on important factors in the pursuit of environmental journalism in an effort to support sustainability communication. In this context, sustainability communication is defined as environmental journalism by journalists who post environmental news continuously.

Factors That Support Journalists' Sustainability Communication

The power of journalism in Indonesia could influence policymakers and ignite public action to solve social problems; hence, there is a need to engage journalists to conduct sustainability communication. This is especially true when journalists perceive that they could contribute to mitigating environmental problems through active and continuous environmental journalism within the context of Indonesia (Agustina, 2017). Through journalists' active role in sustainability communication, it is expected that the issue of environmental problems would become mainstream, get more attention and engagement, and consequently be eradicated via practical actions. Journalists who commit to environmental journalism were found to be stimulated by four factors, which are 'exposure', 'passion', 'knowledge', and 'network'. The factors function in a sequence and, afterwards, change to complement each other. The informants' or journalists' descriptions of the four factors are further elaborated below.

Exposure to Environmental Problems

All of the journalists defined exposure as a continuous interaction with environmental problems. It is the first factor that could switch the journalists' attention from not being aware of environmental issues to directly working in environmental journalism. It is also an incremental factor for the journalists to continue doing sustainability communication. This conclusion is abstracted from all of the journalists' descriptions of how they could keep on doing environmental journalism within the context of Indonesia.

Another important factor in continuous environmental coverage is the journalists' first-hand experiences and interactions with environmental problems, such as being hit by floods or landslides. Other than that, activities that promote environmental discourses also play a role, including on-site reporting on problematic environmental events, participation in workshops/training/seminars/competitions/discussions, or being involved in environmental communities and activities and accessing media and information on environmental issues, such as environmental news coverage, data, or environmental updates, and also having friends and news sources within environmental journalism areas, such as experts, society, and community.

'Exposure' is defined as journalists' direct experience related to environmental problems. It is the turning point for the journalists to enter environmental journalism and ignite their passion. When journalists finally develop awareness and passion, they move into the journalism process. The role of exposure changes when journalists start working in environmental journalism. The role of exposure is to maintain journalists' passion, knowledge, and networking in the field of environmental journalism.

Exposure to environmental issues in Indonesia could be a memorable experience, which was either pleasant or dramatic and sad. The journalists' experiences were mentioned as turning points that made them finally embrace environmental journalism. A journalist described how he/she had been writing environmental news for the last 10 years after winning a fellowship to visit Raja Ampat, covering the Conference of the Parties (COP) in Peru, and being chased and/or washed away by a flood of landslides, among others. Other events that had also ignited journalists' interests in finally working in environmental communication for around 10 years were the experience of losing significant agriculture and/or fishery harvests because of acid rain, or experiencing a tsunami, drought, or landslides that caused the loss of family members, housing, and important documents and materials. All of these experiences prompted the Indonesian journalists to work in sustainability communication for the first time.

After writing about environmental journalism for years, the journalists still needed to experience environmental problems to recharge their passion and knowledge and expand their experience and network. Within the network of environmental journalism, journalists usually have information on fellowship recruitment to work in environmental journalism in a remote area or another country or calls for environmental news proposals, programmes, or submissions to gain new experiences in environmental journalism.

A senior Indonesian journalist, Adhi, described how he needs direct exposure to problematic environmental events at least once a year to recharge his passion for doing environmental journalism, expand his network, and update information or data on environmental problems. Another unique exposure mentioned by journalists is the emotional thrill of success of the news they wrote to influence decision-makers. Surjadi explained that he was thrilled that his research, focus, and investigative news on reducing toxic waste finally encouraged the government to stop importing toxic waste in the late 1990s.

Exposure, therefore, plays two roles for individual journalists: the first as a turning point in their journalistic career, and the second as an enabling factor that maintains their passion for sustainability communication.

Passion in Environmental Communication

Passion is the feeling that comes after the journalists directly experience environmental problems. This feeling could be thrilling or sad, but both trigger or motivate individuals to do more to have the experience again or to hinder it. All journalists perceive that passion is important for them, as the role of passion is to overcome all the challenges that they may incur from doing environmental journalism. It also ignites the spirit of journalists to do more sustainability communication. The journalists also explained that exposure is not the only reason for passion. Passion could also emerge from knowledge. It is easier for journalists who have an educational background related to environmental issues to ignite their passion by reverting to what they already know. However, experiences, especially direct ones, are admittedly the best reason for many journalists to finally have a passion for doing environmental journalism.

'Passion' is the driving factor for journalists to keep on doing environmental journalism. However, individual passion is unstable. It can sometimes be strong or weak. Therefore, it is important to understand how to maintain journalistic passion, so that journalists always have a passion for doing environmental journalism. Elements that can maintain journalistic passion are knowledge, network, and routine exposure to environmental problems. Journalists described how passion will grow over time. The more experience in doing environmental journalism, the deeper and stronger the knowledge and network that maintain passion.

Senior journalists described how passion also relates to the length of time spent reporting environmental problems. Journalists who claim to be environmental journalists usually have more than 3 years' experience in environmental journalism. In Indonesia, the probationary period for a young journalist is around 1 year of training, and 1 year of reporting for different desks. After 2 years, the journalist could be assigned to a certain desk. If the journalist's passion is for environmental themes, he/she could be writing ecological news from various perspectives from a variety of desks.

Unlike exposure, which has to change roles with individual journalists, the role of passion stays the same. Whether it is the first time or during subsequent events, passion always plays the role of driving journalists to do more than merely gather information and data to write about environmental issues. Passion encourages critical thinking in environmental journalism and inspires journalists to make extra efforts to understand the issues that they want to write about. With passion, journalists can do preliminary research; dig for more data;

cross-check information from stakeholders such as experts, the community, and the government in power; and then analyse it within the context of economic, societal, and environmental sustainability presented in news. Furthermore, journalists can plan a step-by-step sequence of news stories to present certain issues to the public, to spark interest and avoid confusion.

Journalists' passion raises their interest, concerns, and sensitivities and prompts attitudes and behaviours in response to environmental problems. These attributes are revealed during their long-term occupations as environmental journalists. Journalists with long-term passion build resilience, commitment, experience, and cultivation of knowledge and skill. The passion of a senior environmental journalist is also usually used to recruit junior journalists to be environmental journalists. He/she could also conduct training and become a resource person, not only for junior journalists but also for others who need insights on certain environmental issues.

Journalists' passion varies based on the kind of information, knowledge, and network they are associated with. The wide scope of environmental problems and environmental issues gives more opportunity to journalists to expand their passion and knowledge on certain topics such as biodiversity, climate change, agriculture, forestry, and birds. It is not easy for Indonesian journalists to master all of the environmental problems owing to the wide and diverse nature of the discourses.

Knowledge in Environmental Communication

Knowledge of environmental problems is identified as an important factor for journalists to pursue sustainability communication through journalism. Knowledgeable journalists report quality news, an idea that was confirmed by all the journalists involved in this study. Environmental problems are complex, and thus news writing on the topics is not an easy task without proper knowledge. Previous scholars have described them as difficult, complex, and sensitive issues, which make them less popular in society. A similar situation is observed in Indonesia, especially in Indonesian television coverage (Agustina, 2020). According to Daeng, journalists need more knowledge not only about the issues, but also about how to convey the message of environmental problems to the public. Often, environmental issues cannot be presented as a one-off news post, but need a series of items.

Journalists who have experienced a tsunami would find it easy to write reports about it based on their experience with it. However, how about news reports on other issues? Journalists need to have more knowledge that enables them to do environmental journalism that is not just based on direct experience. Journalists need to have direct exposure to knowledge as well as experience. More reading and preliminary research in the environmental communication field seem to be essential before writing an environmental news story. Hines et al. (1987) stated that one of the indicators of responsible environmental behaviour is having

TABLE 10.1 Journalists' activities to gain knowledge of environmental problems

No.	*Journalists' activities*
1	Attend seminars, training, workshops on journalism, environmental journalism, and environmental knowledge sessions as participants
2	Become a resources person for seminars, training, and workshops related to the environment, journalism, and environmental journalism topics
3	Become an environmental advocate/environmental consultant/judge on environmental events
4	Join a variety of associations related to the environmental field and become a member of professional journalism
5	Carry out small studies on issues that are going to be covered as news
6	Participate in environmental writing competitions at national and international levels
7	Do site visit reporting/field activities
8	Become an activist in the environmental community

knowledge of the topic. Therefore, journalists who wish to work in environmental journalism for the public's benefit must have knowledge of the matter so that individuals – either journalists or readers – have comprehensive information on the environmental problems being reported.

Knowledge also enhances the quality of journalistic writing, conveying the message to the public so that environmental awareness is developed. There are several ways of gaining knowledge on environmental journalism, besides experiencing it directly, as described in Table 10.1.

Understanding environmental issues by engaging in the activities listed in Table 10.1 is a way for journalists to develop the essential skills to understand the issue as well as critical thinking in order to make intelligent decisions on what to write and to inform the public.

Network in Environmental Communication

To maintain and strengthen the experience, passion, and knowledge of journalists, they need a network of environmental journalism. Environmental journalists' networks include their friends; professional environmental journalist associations (consisting of peers, seniors, and juniors); and news sources that include academicians, political leaders, government institutions/departments related to environmental problems, officials, business owners, industries, environmental non-government organisations, NGOs in general, environmental institutions, opinion leaders, celebrities, and others.

All the informants said that their network is very important. It has many functions to support their professionalism and passion or motivation. For juniors in environmental problems, they found the network very useful because they have mentors to guide them on how to write good coverage of environmental problems. They also have access to data that are already available based on problems/issues covered by the seniors to support issues they want to cover in a shorter time. In general, networks and friends in network become worthy resources to ease journalistic work. Members of a network could access important data that are hard to come by, hence shortening the time to gather the data alone. Networks bring with them a higher level of knowledge and experiences that translate into a wider arena of news coverage. Members of the network could have access to information on scholarships or fellowships at national or international levels. Networks also provide funding to journalists who win competitions for writing environmental journalism pieces.

In general, the network and friends in the network not only maintain and sustain journalists' passion but also help grow journalists' passion to do environmental journalism. They also accelerate journalists' understanding of how to comprehend the big picture of environmental issues. The network facilitates journalists having a positive perception of environmental issues and eases the thinking process to make sense of it. A TV reporter in Makassar (M2) views the network as a channel where she can post concerns and questions, receive confirmation on data, and give and get information on diverse matters related to advancing the capacity of journalism on environmental issues, or sustainability communication. In addition, the network allows access to recharge on other issues in broadcasting media. It is the support that journalists look for in the context of challenging the government in power and media owners' political and economic orientation.

There are a variety of Indonesian environmental journalists' networks. A specific one is the SIEJ. The PWI (Persatuan Wartawan Indonesia, Indonesian Journalists Union) environmental journalists' divisions and AJI (Association of Journalists Indonesia) are among the generalist networks related to climate change, climate leadership, and so forth.

Conclusion

There are four essential factors that encourage Indonesian journalists to pursue sustainability communication through environmental journalism. The first factor is 'exposure', which acknowledges a journalist's experience as a turning point when deciding to do environmental journalism and ignites the second factor, which is 'passion'. Based on a journalist's passion to conduct environmental journalism, he/she needs support from 'knowledge' and 'network'. All of the factors are revealed in a sequence the first time, and, once in place, complement each other, as described in Figure 10.1.

This chapter concludes that journalists should be given direct exposure to events relating to environmental problems to ignite their passion to do sustainability communication. Once the passion exists, journalists should gain knowledge and join a network of environmental groups or associations to sustain the passion. Knowing that these four factors are first revealed in sequence, there is

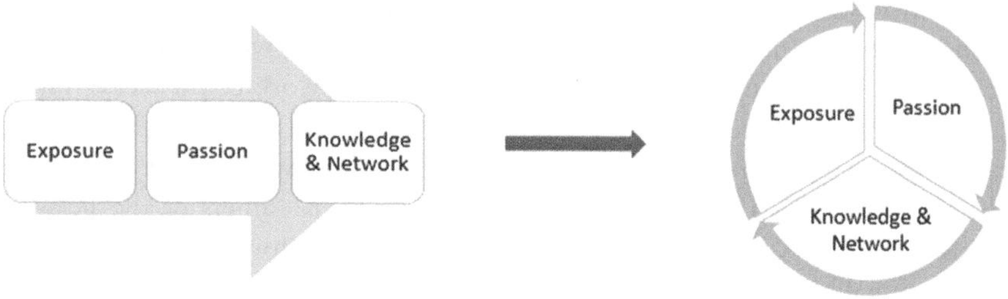

FIGURE 10.1 Important factors for journalists engaging in sustainability communication through journalism work; the factors in sequence then complement each other.

a need to involve more journalists to cover events concerning environmental issues to engage more journalists in environmental journalism. To pursue sustainability communication, environmental journalists need to be more active in exploring their network, which offers more information, opportunities, and activities to raise the capacity of its members.

References

Agustina, A. (2017). *Perceptions and roles of Indonesian journalists in covering environmental problems in Indonesia* [PhD Thesis, Universiti Sains Malaysia]. http://eprints.usm.my/48185/1/PERCEPTIONS%20AND%20ROLES%20OF.pdf%20cut.pdf

Agustina, A. (2020). Environmental communication in Indonesian television news coverage. *Aspiration Journal*, *1*(1), 56–72.

Ambardi, K., et al. (2014). *Country report, mapping digital media*. Indonesia: Open Society Foundations. Accessed from https://www.opensocietyfoundations.org/uploads/9e564507-54b4-40ef-929b-5cf8885adebe/mapping-digital-media-indonesia-20140326.pdf

Dhyaksa, A. (2017, November 4). Karakteristik akun Youtube di Indonesia. *Sci-Tech, Lokadata.id*. Accessed May 29 2022. Accessed from https://today.line.me/id/v2/article/p0LG57

Donsbach, W. (2004). Psychology of news decisions: Factors behind journalists professional behavior. *Journalism Studies*, *5*(2), 131–157.

Fischer, D., et al. (2016). Sustainability communication. *Sustainability Science: An Introduction* (pp. 139–149). Cham: Springer.

Hines, J. M., Hungerford, H. R., & Tomera, A. N. (1987). Analysis and Synthesis of Research on Responsible Environmental Behavior: A Meta-Analysis. *Journal of Environmental Education*, *18*(2), 1–8. http://dx.doi.org/10.1080/00958964.1987.9943482

Lim, M. (2011). *@crossroads: Democratization & corporatization of media in Indonesia*. Research Report. Tempe, AZ: Participatory Media Lab at Arizona State University & Ford Foundation. Accessed from https://cspo.org/legacy/library/1205091216F71753673BI_lib_IndonesiaMedia.pdf

Lim, M. (2012). The league of thirteen: Media concentration in Indonesia. Research Report, Tempe, AZ: Participatory Media Lab at Arizona State University & Ford

Foundation. Accessed from https://www.academia.edu/7282028/Lim_M._2012_The_League_of_Thirteen_Media_Concentration_in_Indonesia.

MacAndrews, C. (1994). Politics of the environment in Indonesia. *Asian Survey*, *34*(4), 369–380.

Mohamad Saifudin, M. S. (2017). Media and environmental non–government organizations (ENGOs) Roles in environmental sustainability communication in Malaysia. *Discourse and Communication for Sustainable Education*, *8*(1), 90–101.

Tesalonica. (2020). Jumlah influencer Indonesia meningkat di tengah pandemi. *tek.id*. Accessed May 31 2022. Accessed from https://www.tek.id/culture/jumlah-influencer-indonesia-meningkat-di-tengah-pandemi-b1ZVp9jeZ

11

COMMUNITY RADIO AND CLIMATE CHANGE IN INDONESIA

Emilia Bassar and Impron

Introduction: Climate Change Issues and Community Radio

The United Nations Framework Convention on Climate Change (UNFCCC) established an international environmental treaty to combat climate change as a statement of determination to protect the climate system for present and future generations. Indonesia signed the UNFCCC in 1992 and ratified it in August 1994 into Law (UU) No. 6/1994 on Ratification of the UNFCCC.

The UNFCCC contains 26 Articles, with the main objective written in Article 2: "stabilisation of greenhouse gas concentrations in the atmosphere at a level that would prevent dangerous anthropogenic interference with the climate system". One of the principles of the UNFCCC is that the parties must act to protect the climate system in accordance with the principle of "common but differentiated responsibilities and respective capabilities".

Based on Article 4 of the UNFCCC, all parties make a general commitment to address climate change through two strategies, namely (1) climate change mitigation and (2) adaptation to climate change impacts. Various modalities, among others, have been carried out through the development of climate policies in development and cooperation for scientific, technological, and capacity building, and so on.

Article 6 of the UNFCCC on Education, Training and Public Awareness is a crucial part of managing climate change and is regularly on the agenda of the Conferences of the Parties (COP). At COP 7 2001 in Marrakech, Morocco, the Marrakech Accord was adopted; it includes a Capacity Building Framework for Developing Countries (2/CP.7), in which 15 priority areas have been identified. Discussion of Article 6 of the UNFCCC at COP 8 in New Delhi, India, in 2002 resulted in the New Delhi Work Program (NDWP), which is multilateral and bilateral cooperation to support activities related to the implementation

DOI: 10.4324/9781003309642-11

of Article 6 of the UNFCCC and its work plan. The elements of Article 6 of the UNFCCC are education, training, public awareness, public access to information, public participation, and international cooperation. At COP 21 2015, in Paris, the UNFCCC adopted the term "action for climate empowerment" (ACE) as the name of Article 6 of the UNFCCC. This term, ACE, gives legitimacy to Article 6 of the UNFCCC as the basic framework for capacity building for climate change.

In 2012, the World Association of Community Radio Broadcasters (AMARC) submitted a pledge to UNFCCC on "Engaging Community Radio on Climate Change Adaptation, Research and Advocacy". AMARC positions itself as a partner to further engage community radio from Africa, Asia, the Caribbean and Central America, Europe, North America, Pacific/Oceania, the polar regions, and South America, with target groups of policy makers, communities, and practitioners (UNFCCC, 2012).

In 2015, UFCCC adopted the Nairobi Work Programme (NWP), which strives to assist all parties – in particular developing countries, including the least developed countries and small island developing states – to improve their understanding and assessment of impacts, vulnerability, and adaptation, and to make informed decisions on practical adaptation actions and measures to respond to climate change on a sound, scientific, technical, and socioeconomic basis, taking into account current and future climate change and variability. Under the NWP, community radio was regarded as a knowledge tool for responding to climate change.

Since then, community radio has been utilised quite extensively in some countries, such as India, Nepal, Guatemala (Basnet, 2021), and Bangladesh (BNNRC & COAST, 2020). In many parts of Africa, radio broadcasts in local languages are a lifeline to the community, and it was regarded as an important medium for raising public awareness on climate change issues in the community (UNFCCC, 2016), for climate change adaptation strategy and action (Cahill, 2019), and for climate change education and resilience (Okaka, 2015). In an advanced country such as Canada, drawing on the resources and broadcasting capacity of community radio – for climate changes issues – was also considered a purposeful choice. Community radio has the potential to buttress community interest and involvement in climate change action (Abdulai et al., 2021).

In discussing the role of community radio and climate change, it will be very helpful if it is associated with ACE elements. This chapter primarily deals with two elements of ACE – that is, public awareness and public access to information. Public awareness aims to reach out and encourage immediate action by the public in the climate crisis. Community awareness activities also aim to support climate-friendly policies, provide opportunities for communities to influence decisions in a democratic process, and encourage changes in people's behaviour through various media (Bassar & Impron, 2014).

Public access to information is an effort to facilitate public access to data and information by providing information about climate change initiatives, policies, and results of actions needed by stakeholders to understand, deal with, and respond to climate change. It is important to facilitate public access to information about climate change initiatives, policies, and results so that it is widely known (Impron, Bassar, & Sugiarto, 2021).

Community Radio Characteristics and Functions

Louie Tabing (as quoted by Pavarala & Malik, 2007) defines community radio as "one that is operated in the community, for the community, about the community and by the community". In the book by Fraser and Estrada titled *Community Radio Handbook* (2001; published by UNESCO), it is explained that the distinctive focus of community radio is to make the public the main actor, through public involvement in all aspects of programme management and production, and presentation to encourage development and social progress in the community. This is in accordance with the basic principles of community radio stated in the Declaration of the World Association of Community Radio Broadcasters/ Association Mondiale Des Radiodiffuseurs Communautaires (AMARC) in 1988, as stated in Fraser and Estrada (2001:3), namely:

> "Community radio responds to the needs of the community it serves, contributing to its development within progressive perspectives in favour of social change. Community radio strives to democratise communication through community participation in different forms in accordance with each specific social context."

AMARC states that "community media, especially community radio, are very important for developing countries around the world, where community radio is sometimes the main means of communication to convey news and information" (Howley, 2005: 14). Regarding the relationship with community members, Solervicens (2008: 6) adds that: "community radio has the distinctive capacity to provide pluralistic and participatory communication that is receptive to the need for expression from the social and cultural sectors as compared to other media, particularly commercialised media".

The characteristics of community radio established by the community and involving community members in the management and design of programme production allow them to independently determine information services for the development and progress of their community. As explained by Fraser and Estrada (2001:4),

> "A community radio station is characterised by its ownership and programming and the community it is authorised to serve. It is owned and controlled by a non-profit organisation whose structure provides for membership, management, operation, and programming primarily by members of the community at large. Its programming should be based on

community access and participation and should reflect the special interests and needs of the listenership it is licensed to serve."

(at page 4)

The main functions of community radio are detailed by Fraser and Estrada (2001), as follows: (1) reflect and promote local identity, character, and culture; (2) create a diversity of voices and opinions on the air; (3) provide a diversity of programmes and content; (4) encourage open dialogue and democratic process; (5) promote development and social change; (6) promote civil society; (7) promote good governance: (8) encourage participation, sharing of information, and innovation; (9) give voices to the voiceless; (10) provide a social service as a replacement for the telephone; (11) contribute to diversity in broadcasting ownership; and (12) contribute to the development of human resources for the broadcasting industry.

The characteristics and functions of community radio allow the use of community radio as a medium for the community to get easy access to climate change information, how to deal with climate change, use local wisdom in dealing with climate change, and others.

Community Radio and Climate Change Issues in Indonesia

Community radio in Indonesia is legally recognised in Law No. 32/2002 on Broadcasting. As the number of community radio stations in Indonesia was increasing, this prompted community radio activists to establish the Jaringan Radio Komunitas Indonesia (JRKI)/Indonesian Community Radio Network (ICRN) in 2002. JRKI has 18 community radio networks in 18 provinces, which means it covers more than half (53%) of the provinces in Indonesia – that is, 18 out of 34.

Apart from regional-based associations at the national level (JRKI) and at provincial level (JRK Province), there are also other forms of community radio associations based on sectoral typologies. For example, the Jaringan Radio Suara Petani (JRSP)/Voice of Farmers Radio Network is an association of farmer and fisherman community radio stations in West Java and Lampung Provinces, and Forum Radio Kampus Bandung (FRKB)/Bandung Campus Radio Forum is a forum uniting all the campus radio stations in Bandung Raya, West Java.

Until the end of 2005, there were no accurate data on the number of community radio stations in Indonesia. The reason was, partly, because research on the development of community radio tended to be sporadic and was carried out based on the sectoral interests of the research provider, not holistically. According to an estimate by the Combine Resource Institution (CRI) in 2005, there were 451 community radio stations in Indonesia, with the largest number located on the island of Java (Masduki, 2004).

Generally, community radio in Indonesia focuses more on broadcasting local issues of concern to local people. Broadcasts are packaged as various talk shows between residents discussing special topics or as entertainment programmes, broadcasting traditional stories or songs of various genres. Community radio is also used as a medium for sharing contemporary or current information that is considered a national or even global issue. An example is environmental issues, including climate change.

The Indonesian Agency for Meteorological, Climatological and Geophysics (BMKG), The Indonesia Climate Change Trust Fund (ICCTF), and the Indonesian Institute of Sciences (LIPI) took the initiative in 2010 to utilise community radio as a tool for collective learning to increase climate change adaptation capacity among members of the farming and fishing communities. There were at least six reasons for choosing radio as a socialisation medium that can convey information related to climate change accurately and quickly, namely it is the choice of farmers and fishermen, mobile, interactive, adaptive and responsive, effective and efficient, and rich in local content. Six radio stations were selected to disseminate climate change in five regions, namely Indramayu, West Java; North Jakarta, DKI Jakarta; Batu, East Java; Serdang Bedagai, North Sumatra; and Bau-Bau, Southeast Sulawesi (Abdurrarahim, 2012).

Another example concerning climate change in the radio community was found in Gema Solidaritas community radio in Ketapang, West Kalimantan Province. This radio station made a broadcast programme that could influence people's mindsets to respond to climate change issues. This programme was a USAID IFACS initiative that concerned environmental issues and climate change (Amiruddin, 2017).

Based on the results of media monitoring for the period 2020–2021 on community radio and climate change, many climate change issues were related to flood management and forest conservation (see Table 11.1). Meanwhile, posts on social media were more diverse, namely about humans and nature, commemorating Earth Day, environmental sustainability, blue carbon, and composting (see Table 11.2). The use of social media by community radio is mostly done by community radio stations in universities. In addition to various topics, the visual material posted was also more interesting and creative (see Figure 11.1). This could be owing to the involvement of students of community radio management who have skills in managing social media and good knowledge of environmental issues and climate change.

Meanwhile, community radio in rural and coastal areas is usually managed by local residents who have limited knowledge of the field of high technology and the use of social media to reach a wider audience and more followers.

There are many climate change issues that could potentially be discussed on community radio. Items to increasing public awareness of agricultural issues due to climate change, forest fires, rising sea temperatures, and animal protection could be broadcast on community radio in local languages, with local issues and local wisdom. For example, Bassar (2015) found that one of the mainstay programmes of Ruyuk FM community radio was *Conservation and Livelihoods*, which dealt with the importance of preserving nature and biodiversity in

TABLE 11.1 Examples of news on community radio and climate change in 2020–2021

No.	*Media Name*	*News Header*	*Date*
1.	radiogamasi.com	*Komunitas Jurnalis Berbagi Gandeng Berkah Box Bantu Korban Banjir Makassar* (Journalist Community Shares Blessing Box to Help Makassar Flood Victims)	22/12/2020
2.	makassar.terkini.id	*Siaga Bencana, Warga Gowa Lakukan Simulasi Peringatan Dini dan Evakuasi Banjir* (Disaster Preparedness, Gowa Residents Conduct Flood Early Warning and Evacuation Simulation)	22/01/2020
3.	radardepok.com	*Komunitas Green Campboun Depok: Punya Hutan Percontohan, Jaga Kelestarian Pesanggrahan (1)* (Green Campboun Depok Community: Have a Pilot Forest, Maintain the Presence of Pesanggrahan (1))	11/03/2020
4.	kronikkaltim.com	*LKK Kutim beserta Organisasi Sayap Gandeng Komunitas Radio Galang Dana Korban Banjir* (LKK Kutim and Wings Organizations Collaborate with Community Radio to Raise Funds for Flood Victims)	25/05/2021
5.	rdk.fidkom.uin jakarta	*Kolaborasi RDK FM dan ACT untuk Galang Dana Korban Bencana Alam di Indonesia* (RDK FM and ACT Collaboration to Raise Funds for Natural Disaster Victims in Indonesia)	21/01/2021

Mandalamekar Village, Tasikmalaya Regency, West Java, Indonesia. In this programme, the community was invited to create nurseries, planting and maintaining trees in the forest.

This was in accordance with the mission and objectives of the Mitra Alam Munggaran (MAM) self-help group in Mandalamekar Village. This community group cares about the environment, having local wisdom known as *leuweung nganteng kaca nunggal* (sustainable forest mirrors life). Tata Sumitra, an MAM activist, explained that he was informed that global warming has caused climate change that is detrimental to farmers, such as the lack of water availability for agricultural land. To deal with the impacts of climate change, MAM has pioneered the replanting of damaged forests on the edge of the village. The community believed that planting trees in the forest could revive the source of water they needed for agricultural land and household needs in the form of clean water.

Reforestation activities were not only broadcast on air but also coupled with off-air activities in the form of visits by radio activists, with listeners and

TABLE 11.2 Examples of content on social media about community radio and climate change in 2020–2021

No.	*Media Name*	*Caption*	*Date*
1.	Instagram Radio Unpad	[A Virtual Project Hosted By Antropologi Budaya UGM] AnthroFest 2020 "(re)becoming human, (re) connecting lives" We're inviting you to join our virtual celebration Save the dates and let's celebrate humanity with AnthroFest 2020! Official Line: @radiounpad (use @) Partnership & Media Partner: radio@unpad.ac.id Instagram & Twitter: @radiounpad Website: radio.unpad.ac.id #RadioUnpad #Sivitas	09/08/2020
2.	Instagram rdk_fm Radio Dakwah dan Komunikasi	*Selamat Hari Bumi* (Happy Earth Day) *Yuk, Balamuda bergerak bersama mengampanyekan Hari Bumi sebagai wujud peduli lingkungan. Balamuda juga bisa share ucapan selamat Hari Bumi di media sosial masing-masing loh* (Come on, Balamuda move together to campaign for Earth Day as a form of caring for the environment. Balamuda can also share congratulations on Earth Day on their respective social media) *Mari menjaga kelestarian bumi dan mencintainya seperti yang dilakukan oleh nenek moyang kita!* (Let's preserve the earth and love it like our ancestors did!) #RDKFM #radiouinjkt #haribumisedunia	22/04/2020
3.	Podcast Cleanomic Radio	Cleanomic Radio Podcast @cleanomictweets *Membahas tentang cerita orang-orang, kegiatan, inisiasi, komunitas, & gerakan yang peduli kelestarian lingkungan. Topiknya beragam mulai dari karir yg mendukung kelestarian lingkungan, karbon biru, kompos, & masih banyak lagi* (Discussing stories of people, activities, initiations, communities, & movements that care about environmental sustainability. The topics range from careers supporting environmental sustainability, blue carbon, composting, and many more)	30/04/2020
4.	Instagram Radio Unpad	*3 Juni 2021 adalah Hari Bersepeda se-Dunia. Mari mulai biasakan bersepeda dan jangan lupa untuk selalu mematuhi aturan lalu lintas* (3 June 2021 is World Cycling Day. Let's get used to cycling and don't forget to always obey traffic rules) http://unpad.ac.id/twibbon	03/06/2021

(*Continued*)

TABLE 11.2 Continued

No.	*Media Name*	*Caption*	*Date*
5.	107.8 FM Atma Jaya Radio	*Hallo Sobat Atma!* (Hallo Sobat Atma!) *Tahukah kamu bahwa daerah terbasah di dunia, ternyata ada di Indonesia! Daerah tersebut berada di Kecamatan Tembagapura, Kabupaten Mimika* *Hal ini pun berangkat dari penelitian Tim BMKG (Badan Meteorologi, Klimatologi, dan Geofisika) daerah setempat pada Oktober 2018 lalu yang melakukan penelitian bersama Tim Environmental PT Freeport Indonesia (PTFI), yang dilakukan khususnya pada jalan tambang Mile 50 Post, Tembagapura* (Did you know that one of the wettest areas in the world is actually in Indonesia! The area is located in Tembagapura District, Mimika Regency This was the result of research by the local team of BMKG (Meteorology, Climatology and Geophysics Agency) in October 2018 in collaboration with PT Freeport Indonesia (PTFI) Environmental Team, which was carried out especially on the Mile 50 Post mining road, Tembagapura) *Selama 5 tahun terakhir, daerah ini memang telah diguyur hujan selama 329 hari setahun atau 12,143 milimeter sepanjang tahun. Catatan ini sekaligus melebihi rekor sebelumnya yang dipegang oleh India. Hingga pada akhirnya, BMKG Timika bersama dengan Tim PTFI mendaftarkan daerah Mile 50 ini sebagai daerah paling basah di dunia ke World Meteorological Organization (WMO) yang merupakan badan meteorologi dunia* (Over the past 5 years, this area has indeed been rained for 329 days a year or 12,143 millimeters throughout the year. This record also exceeds the previous record held by India. In the end, BMKG Timika together with the PTFI Team registered the Mile 50 area as the wettest area in the world to the World Meteorological Organization (WMO) #Ajrfunfact #Radiokomunitas #sobatterbaikmu #wonderfulindonesia #Mimika	15/11/2021
6.	Mas Bagus PHK	*Kegiatan reboisasi komunitas Radio Jago Kediri embali koramil Mojo dengan tujuan penghijaun embali untuk melestarikan alam sekitar, mencegah banjir, dll* (The reforestation activity of the Radio Jago Kediri community with the Mojo Koramil with the aim of reforesting to preserve the surrounding nature, prevent flooding, etc.)	28/12/2021

FIGURE 11.1 Examples of content on social media of a campus-based community radio station published on Twitter (Radio Unpad, 2020).

villagers, to the protected forest of Karang Soak to clear weeds in the forest. This activity also involved junior high school students and the Community of Women Environmental Activists (KPPL), which was formed by MAM in 2008. The night after the activities took place, news and stories about the visit to the Karang Soak Forest were broadcast by Radio Ruyuk FM.

Radio Ruyuk FM raised the issue of community interest regarding the water crisis due to uncontrolled logging in order to change the behaviour of the community so as not to cut down trees in the forest, especially in protected forests and community forests. The participatory approach of involving listeners in the activities of Radio Ruyuk FM made listeners not only listen but also take action for environmental conservation, for example in the management of spring water resources (see Figure 11.2).

The Ruyuk FM radio broadcast on forest conservation showed its impact in the dry season in 2011, when villagers still had water supplies for agricultural land sourced from springs in the forest of Mandalamekar Village. According to Yana Noviadi, head of Mandalamekar Village, during the dry season of 2011, farmers were able to irrigate 80% of their rice fields, while, in other villages, the rice fields dried up owing to the dry season. Previously, every dry season would bring a water shortage for agricultural land, causing conflicts between the villagers. Villagers also found it was difficult to get clean water for household needs (Bassar, 2015).

Another example relates to the background of the establishment of the Rakita FM community radio in Bandung City, West Java. The community was faced with poor waste management that caused flooding. Therefore, one of the flagship programmes of Radio Rakita FM was *Green Generation Time*, which raises

FIGURE 11.2 Spring water management found in the forest of Mandalamekar Village, West Jawa, Indonesia (Documentation: Emilia Bassar, 2/7/2011).

issues about global warming, the greenhouse effect, waste management in the surrounding environment, and the like. The *Green Generation Time* programme was in line with the initial goal of establishing radio to protect the environment so that the environment becomes beautiful, clean, and healthy and is not affected by disasters such as floods. Broadcasters conveyed various information about the environment sourced from the internet and newspapers. The radio manager also invited a resource person and gave listeners the opportunity to interact with resource people about environmental issues, climate change, or global warming (Bassar, 2015).

As shown by the results of media monitoring for the period 2020–2021, not many community radio stations have specific programmes on climate change; instead, climate change was embedded as an environmental problem. This condition raises the need for capacity building to improve the role of community radio in dealing with climate change issues. Increased knowledge and understanding of climate change can be achieved through capacity building for community radio managers through various programmes. Capacity building can be implemented in collaboration with the government, companies, universities/scientists, and non-governmental organisations (NGOs). By acquiring better knowledge, community radio would better serve as a communication medium to increase community awareness and access to information about climate change.

It is also important to notice some common problems in the operation of community radio. Based on interviews with the managers of 20 community radio stations who were members of JRKI in Central Java, it was found community radio had problems with management, funding, community participation, and support from third parties (Sanjaya & Pandiangan, 2019). In an earlier study, Masduki (2004) pointed out two of the biggest problems with community radio, namely licensing and improving broadcast quality so that it is able to build the participation of the targeted community. Support from external forces such as NGO activists, academics, and private and government institutions should be oriented to address these common problems in order to empower community radio as a medium for disseminating information, knowledge, and skills for climate action.

Increasing Community Awareness and Access to Information through Community Radio

Capacity Building: Public Awareness

The capacity building activities of the community awareness element have the main objective of reaching out and encouraging immediate action by the community in the climate crisis. Public awareness activities also aim to support climate-friendly policies, provide opportunities for communities to influence decisions in a democratic process, and encourage changes in people's behaviour through various media (Bassar & Impron, 2014).

Cooperation between stakeholders in the area of public awareness is needed to bring climate change issues to the public's attention, increase community capacity in handling climate change, and encourage community action with contributions and actions to overcome climate change (Impron, Bassar, & Sugiarto, 2021).

UNESCO, in a book entitled *Integrating Action for Climate Empowerment into Nationally Determined Contributions – A Short Guide for Countries* (2020: 7), provided a list of the six elements of public awareness, quoted as the following:

1. Cooperate in, promote, facilitate, develop, implement inclusive public awareness programmes on climate change and its effects at the local, regional and national levels as appropriate;
2. Create key messages that make it simple and fun for people to take climate action and develop and implement action campaigns to encourage people to do small, easy things that can lead to lasting change, such as reducing their personal water or electricity use;
3. Encourage, by means of popular media, personal action and contributions to address climate change, support climate-friendly policies and foster behavioural changes, noting the important role that social media and digital platforms and strategies can play in this context;

4. Collaborate with civil society and private sector partners to ensure wide dissemination of public awareness-raising activities and limit the spread of "fake news" and disinformation campaigns;
5. Conduct surveys to establish a baseline of public awareness, which can serve as a basis for further work and support the monitoring of the impact of activities;
6. Recognize and engage youth as knowledge holders and support the participation of youth as agents of change and/or trainers in public awareness campaigns.

By increasing public awareness, community radio can create broadcasting strategies related to climate change issues by determining the goals or objectives of the broadcast programme, identifying target listeners and supporters, compiling main weekly or monthly themes that are detailed further to broadcast topics, developing more creative programmes (on air, off air, or hybrid), and measuring the success of the broadcast programme.

Considering that the people involved in community radio are also acting as "journalists" who look for materials to prepare broadcasting materials, therefore, various efforts have been made to increase the capacity of these journalists by providing handbooks, such as *Getting the Message Across. Reporting on Climate Change and Sustainable Development in Asia and the Pacific: A Handbook for Journalists* (UNESCO, 2019). This book provides journalists with practical knowledge on how to report the complex phenomena of climate change and sustainable development and tailor them to specific conditions that are understandable by the local community. The book is available in Indonesian and is accessible online.

Capacity Building: Public Access to Information

Closely linked to building public awareness are the many actions that ensure information on climate change remains transparent and accessible. Public access to information is not just about spreading knowledge; it also questions the broader community-based structures and social forces that shape how knowledge is spread across sectors and across local, national, regional, and international communities. Public access to information also includes feedback and mechanisms that link decision-makers, practitioners, and those directly affected by the adverse impacts of climate change to share their understanding and experiences (UNFCCC, 2022).

Access to information and freedom of expression are human rights guaranteed by Article 19 of the United Nations Declaration of Human Rights and Article 22 of the International Covenant on Civil and Political Rights (ICCPR). The two articles guarantee not only the freedom of the public to convey information and all their ideas, but also the freedom to seek and receive information in various forms and types of media.

Article 19 of the Declaration of Human Rights includes not only the concept of information, but also the concept of communication. On this basis, the public can become creators, users, and distributors (channels) of information. Communication includes action and interaction. Communicating is expressing personal opinions and tendencies, including sharing and exchanging ideas. The concept explains that the circle between communication and information describes the relationship between information and communication components with three democratic governance frameworks, namely transparency, active participation, and responsive/responsible attitude.

UNESCO (2020) states that it is necessary to guarantee public access to relevant information and data about climate change initiative practices, policies, and the results of their actions, so that people can understand and respond to climate change. These efforts should take advantage of new information technologies, taking into account various factors, such as the quality of internet access, literacy, gender, and language issues. Community radio provides an alternative tool for the public to access information with a limited internet infrastructure.

Community radio can provide information to local communities and listeners by broadcasting information – for example, on agriculture, forestry, waste, and climate change adaptation and mitigation. Community members have the opportunity to contribute to managing and developing radio according to their knowledge and abilities. Community radio offers local residents the opportunity to design radio programmes and improve their technical skills.

Community radio provides an opportunity for listeners, including women and youth groups, to gain access to information relevant to their community issues. Solutions to community problems can be communicated through radio broadcasts. However, special strategies or tips are needed to involve the community in radio management.

Conclusion

The BMKG-ICCTF-LIPI took the initiative in 2010 to experiment with six community radio stations as a tool for collective learning among members of the farming and fishing communities to increase their climate change adaptation capacity. The use of community radio was seen as an appropriate communication medium for communities in rural agricultural and coastal areas.

A study in 2011 on two community radio stations in West Java, Indonesia, namely Ruyuk FM and Rakita FM, showed that broadcasts related to environmental issues dominated more than climate change issues. The landscape of community radio broadcasting content has not changed significantly in recent years, as indicated by the results of media monitoring for the period 2020–2021. Community radio seemed to operate in a business-as-usual style, airing content that was more on issues of community affairs, agriculture, forestry, fisheries, waste management, and environmental degradation. Nevertheless, as these issues are relatable to climate, they could provide potential entry points for packaging issues of climate change in the context of handling local environmental and livelihood problems, as well as disaster risks related to climate or weather.

Learning from experiences in other regions, we envisage that capacity building is needed in order to increase the knowledge and understanding of the personnel managing community radio operation of the complex phenomenon of climate change. The effort should synergise collaboration between relevant stakeholders from government agencies, NGOs, companies, and media organisations. The objective of the capacity building is not to alter the best practices already attached to business-as-usual community radio operation, but rather to enrich existing broadcasting with content related to climate issues. This would further empower community radio as a medium for community awareness raising and access to information on climate change issues.

References

Abdulai, A., Chireh, V. K., & Tchoukaleyska, R. (2021). Engaging diverse audiences: the role of community radio in rural climate change knowledge translation. *Journal of Community Engagement and Scholarship, 13*(3), 108–121.

Abdurrahim, A. Y. (2012). Radio sebagai media sosialisasi perubahan iklim. In D. Hidayati & E. Aldrian (Eds.), *Perubahan iklim: upaya peningkatan pengetahuan dan adaptasi petani dan nelayan melalui radio* (pp.131–154). PT Sarana Komunikasi Utama berkerja sama dengan ICCTF, BMKG, dan LIPI. Jakarta, Indonesia.

AMARC. (2007). *Community radio social impact assessment: Removing barriers increasing effectiveness. Challenges, findings, reflections, experiences, lines of action for community radio stakeholders.* Montreal (Quebec), Canada: AMARC Global Evaluation. .

Amiruddin, Z. (2017). Peran radio komunitas Gema Solidaritas sebagai pemberdayaan masyarakat Ketapang. *Jurnal Interaksi, 1*(1), 50–71.

Basnet, S. (2021). Community radios to promote indigenous & ancestral knowledge for climate change adaptation [Paper presentation]. COP26, November 5, 2021. Accessed from https://unfccc.int/sites/default/files/resource/AR_Day_Presentation_1-IGES-ONAMIAP_%28English%29.pdf

Bassar, E. (2015). *Relasi Gender dalam Keterlibatan Perempuan di Radio Komunitas (Studi pada Radio Ruyuk FM dan Radio Rakita FM Di Jawa Barat).* [Unpublished Doctoral Dissertation, University of Gadjah Mada]. Universitas Gadjah Mada

Bassar, E., & Impron. (2014). *Indonesia national strategy on Article 6 of the United Nations Framework Convention on Climate Change (UNFCCC).* DNPI and Japan International Cooperation Agency (JICA). Jakarta, Indonesia.

BNNRC & COAST. (2020). Community climate airwaves through community radio in Bangladesh. BNNRC. Accessed from https://drive.google.com/file/d/1DVJ9bP_ZgBVNx6Pn96wRr6_PkLLTXjsi/view?fbclid=IwAR1GGsh3MMU3iqmtWN4URCdCQ6gR1ZL_08NZ87xva9r8tuXjEgsDyeS0kSw

Cahill, A. (2019, October 14). How local radio boosts climate change adaptation. Norwegian Refugee Council. Accessed from https://www.nrc.no/expert-deployment/2016/2019/radio-programming-boosts-climate-change-adaptation/

Fraser, C., & Estrada, S. R. (2001). *Community radio handbook* (pp. 13–23). Paris: UNESCO.

Howley, K. (2005). *Community media: People, places, and communication technologies.* Cambridge, England: Cambridge University Press.

Impron, Bassar, E., & Sugiarto, Y. (2021). *Development of Indonesia's capacity building roadmap in relation to the implementation of the Indonesia's Nationally Determined Contribution.* Jakarta, Indonesia: KLHK, GIZ, and CPROCOM.

Masduki. (2004). Perkembangan dan problematika radio komunitas di Indonesia. *Jurnal Ilmu Komunikasi, 1*(1), 73–86.

Okaka, W.T. (2015). Effectiveness of community radio in communicating climate change policy for community education and resilience in the East African community. In Paper presentation Conference: 8th World Environmental Education Congress, Gothenburg, Sweden.

Pavarala, V., & Malik, K. K. (2007). *Other voices-the struggle for community radio in India.* New Delhi: Sage Publications India Pvt Ltd.

Radio Unpad [@RadioUnpad]. (2020, December 11). *[A virtual project hosted by Antropologi Budaya UGM] AnthroFest 2020 "(re)becoming human, (re) connecting lives" We're inviting you to join* [Tweet; thumbnail link to article]. Twitter. Accessed from https://twitter.com/RadioUnpad/status/1337372734989295616

Sanjaya, A. R. & Pandiangan, A. (2019). Pergulatan di balik frekuensi: studi deskriptif radio komunitas di Jawa Tengah. *CALATHU: Jurnal Ilmu Komunikasi, 1*(1), 42–51.

Solervicens, M. (Ed.). (2008). *Women's empowerment and good governance through community radio: Best experiences for an action research process.* Montréal: AMARC, World Association of Community Radio Broadcasters.

UNESCO. (2019). Getting the message across. reporting on climate change and sustainable development in Asia and the Pacific: A handbook for journalists. UNESCO Bangkok. Accessed from https://bangkok.unesco.org/content/getting-message-across-reporting-climate-change-and-sustainable-development-asia-and

UNESCO. (2020). Integrating action for climate empowerment into nationally determined contributions: A short guide for countries. *UNESCO Digital Library.* Accessed from https://unesdoc.unesco.org/ark:/48223/pf0000373762

UNFCCC. (2012). Engaging community radio on climate change adaptation, research and advocacy. Accessed from https://www4.unfccc.int/sites/NWPStaging/Pages/item.aspx?ListItemId=23177

UNFCCC. (2016). Accelerating climate action through public awareness. Accessed 13 June 2022. Accessed from https://unfccc.int/news/accelerating-climate-action-through-public-awareness

UNFCCC. (2022). Public participation under Action for Climate Empowerment. Accessed 13 June 2022. Accessed from https://unfccc.int/topics/education-and-outreach/workstreams/public-participation

12

TOWARDS SUSTAINABILITY IN CORPORATE SOCIAL RESPONSIBILITY

Suriati Saad, Jamilah Ahmad, and Mahadevan Krishnan

Introduction: Organisation and Corporate Social Responsibility

Corporate social responsibility (CSR), which incorporates the economic, legal, ethical, and discretionary expectations that society has of organisations over time, has been defined as the social responsibility of a firm (Carroll, 1979). Sustainability, a concept that started in the economics of natural resources but has since expanded in terms of sustainable development and social equality, is the capacity to maintain environmentally, socially, and economically varied systems and processes (McKinsey, 2019).

As organisations grow their networks nationally and internationally, they are required to interact with various communities surrounding the organisation, reflecting that they do not belong to a single stakeholder. Still, a more comprehensive range also indicates a collective responsibility that can benefit the society. Corporate social responsibility serves shareholders' interests and has become an essential field for businesses (Nalband & Kelabi, 2014). With the suggestion that some businesses would claim that CSR is illegal because it violates the owners' fiduciary duty to maximise shareholder profits, Friedman (1970) began to make the case that corporate responsibility hinders a company's ability to maximise profit.

Thanks to the efforts of pressure groups and the rise of "virtuous markets" such as socially responsible investments (SRIs), which have increased pressure on organisations to implement CSR programmes, CSR has gained commercial significance. In any organisation, employee engagement in CSR is an important cornerstone. Previous research demonstrates the benefits of social responsibility regarding employee engagement and the growing need for organisations to take an interest in developing corporate responsibility programmes. Researchers

DOI: 10.4324/9781003309642-12

are beginning to understand that earlier studies of CSR did not consider specific employee outcomes of CSR (Glavas, 2016). The major stakeholders in a company are the employees, who participate in, contribute to, and respond to the CSR policies of the company, particularly those that are directly relevant to them.

Defining Corporate Social Responsibility

There is no absolute definition or one-size-fits-all purpose in defining CSR, making its definition a complicated task (Asongu, 2007; Rexhepi, Kurtishib, & Bexhetic, 2013; Sheehy, 2015). Although the concept of CSR has provided boundaries and debatable legitimacy, most scholars agree that the CSR dimension involves ecology, society, and economic systems with multifaceted nature problems that are highly complex and dynamic. Still, the ubiquity of CSR dialogues and the amount invested in the sectors indicate the importance of accurately defining the term. Without a clear explanation of what it is and what it entails, CSR is perceived as a new dimension, leading businesses to interpret their obligations in relation to social reporting differently (Nehme & Wee, 2008). Companies adopt, refine, and develop their strategies and competitive advantages traversing their institution's environments as they do not operate in an existential vacuum, which creates both barriers and opportunities (Gjølberg, 2009).

A company's commitment to economic and social sustainability, as well as to being a responsible steward of the environment and social landscapes it operates in, is referred to as corporate social responsibility (Ahmad & Crowther, 2013).

In Malaysia, social and environmental dimensions become the critical factors of organisations' commitments to practising CSR. Research suggests that

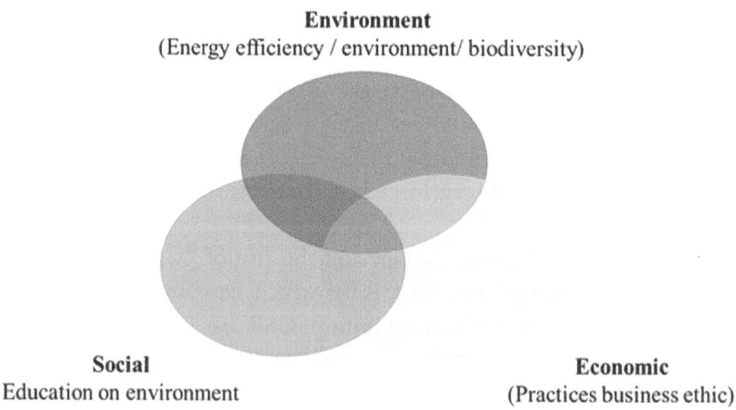

FIGURE 12.1 Environmental, economic, and social dimensions (Ahmad & Crowther, 2013).

companies concerned about their responsibilities towards their workforce, customers, and communities are more financially successful and create a positive reputation with their stakeholders. PETRONAS, in Malaysia, adopted seven dimensions of key CSR areas that overlap with sustainable development goals. The seven CSR areas embedded in PETRONAS's mission and vision have turned the company not only into one of the top oil-and-gas organisations in Malaysia but one that is profitable and often creates an impact on the society through its social engagement activities (PETRONAS Integrated Report, 2020).

CSR's concept and definition could be aligned with the organisation's aim, intention, and strategy in responding to the circumstances of its operations (Van Marrewijk, 2003). The creation and perception of CSR have been dynamic as a result of the influence of cross-cultural norms, multiple country locations, and varied internal and external environment changes. With change in organisations' practices, CSR has changed and grown, manifested itself, and flourished (Carroll, 2008). Different definitions by various authors give different dimensions of CSR (Kamal, 2021), with the meaning evolving and concepts overlapping (Masoud, 2017). Hence, the underlying concept and importance of CSR are very much dependent on the context of the practice as well as the environment surrounding the practice.

CSR and Society

Businesses will need to make significant adjustments to their operations in order to operate in harmony with the social environment as a result of the long-term effects of these social policies, as a strategy to obtain and maintain community acceptance (Mishra, 2019; Musa, Yusuf, & Mcardle, 2013; Du, Bhattacharya, & Sen, 2010; Kulkarni, 2017; Mahmud, Ding, & Hasan, 2021). The issues include the organisation of labour and safety, corporate profitability, economic stability, and the preservation of the environment (Sheehy, 2015). CSR initiatives can and should serve both the needs of society and the bottom line of a business, while enhancing the credibility of its CSR message to generate goodwill (Du, Bhattacharya, & Sen, 2010). According to Nehme and Wee (2008), CSR concepts also provide a pathway for a business to concern itself with the social dimension and social impact.

CSR's role in the society has become substantial and benefits organisations and key stakeholders. CSR encompasses the relationship between a business and the society in which it operates, with CSR characterised as a rich diversity of views on the role of business in society (Asongu, 2007). CSR extends past the company's walls and into the community, involving a broad range of stakeholders, including business partners, suppliers, customers, government organisations, and NGOs (Asongu, 2007; Vazquez, Lanero, & Licandro, 2013; Ahmad & Krishnan, 2017).

In Malaysia, giant telecommunications organisations have often created an impact through their CSR initiatives towards society, such as is shown in Figure 12.3.

FIGURE 12.2 PETRONAS's seven CSR mission and vision areas (PETRONAS Integrated Report, 2020).

DIGI.COM BERHAD
Through their innovation in product and services, they have attracted more than 12.1 million customers in the telecommunication industry. The Telenor Group was one of Forbes' 100 World's Most Innovative Companies in 2014. The firm is famous for creating social-economic developments and assisting companies with technology advancement. DIGI's company helped more than 38,098 school kids and reached out to more than 130 ICT teachers understand online safety methods. Their program also helped children living in poor urban centers (https://sustainability.digi.com.my/)

AXIATA GROUP BERHAD
Axiata is recognized as the telecommunications titan in Asian countries with 290 million subscribers in 10 countries. With control brands such as 'XL' in Indonesia, 'Dialog' in Sri Lanka and 'Celcom' in Malaysia, not to forget the 'edotco' which operates across six countries, they hit an impressive RM19.9 billion in 2015.
The multinational company continues to support its IT CSR programs Kampung Tanpa Wayar (KTW) and Pusat Internet Malaysia. Thanks to their CSR initiatives they got the third place under the Edge Billion Ringgit Club Corporate Awards 2014 for CSR initiatives (https://sustainability.axiata.com/)

FIGURE 12.3 Digi and Axiata telecommunications organisations' CSR initiatives (Digi & Axiata, 2022).

Previous research concluded the importance of CSR to the community, whereby CSR has implications for society and development and can be used as a device to solve community problems. Research carried out by Rosilawati and Ahmad (2019) revealed the successful implementation of CSR initiatives by Aqua Bali Company in Indonesia, which, by engagement with local communities and adapting local wisdom concepts, succeeded in conserving the Ayung River in Indonesia, which is the longest river on the island of Bali. The seriousness of the involvement of Danone-Aqua, Indonesia's largest bottled water company, with the local community has developed trust and long-term relationships between the company and stakeholders. Danone-Aqua's CSR initiatives are classic examples of deeply rooted commitment and engagement between an organisation and society. Likewise, Malaysian corporate organisations should also ensure such an impact through participation and commitment from all stakeholders. In order to do it, we must reveal the history of the difficult and profoundly impactful connection between business and society. CSR encompasses larger categories such as community-driven components, as well as human rights, labour rights, environmental stewardship, and social reporting. CSR refers to a variety of duties in several areas, including the environment, labour, and financial services, and ranging from benevolent to legal. CSR is thought to be the answer to the world's problems of poverty, social marginalisation, and environmental damage (Van Marrewijk, 2003).

Business: Responsibility for the Community

Since CSR is constantly influenced by shifting environmental and cultural factors, there has never been one single definition that encompasses all of its dimensions (Mark-Herbert, Rotter, & Pakseresht, 2018; Aaijaz & Ibrahim, 2012). Economic, legal, ethical, and philanthropic responsibilities are the four main categories delineated in a pyramid model by Carroll (1979). Carroll's CSR model illustrates the operation of the business as being economically profitable, law-abiding, ethical, and socially supportive, adapting for universality (Nalband & Kelabi, 2014). In 1999, revisiting his CSR model created in 1979, Carroll presented the CSR pyramid, a framework for understanding the evolving nature of an organisation's economic, legal, ethical, and philanthropic performances. CSR concepts have endured over a period while continuing to grow in importance and impact (Carroll, 2016). Social responsiveness, social arrangement, public policy, ethics, and stakeholder management have supplanted CSR terms over time.

Hence, the corporate responsibility macro-focus and ethical behaviour are more of micro-focus, and business ethics share a moral code as the tool for achieving ethical behaviour (Fassin, Rossem, & Buelens, 2011). When CSR and ethical principles are implemented and practised coherently, organisations attain economic efficiency, confidence, and cultural identification, leading to competitive companies (Gomez & Manzanares, 2011). According to Migle (2015), organisations invest in socially responsible activities for several reasons. Companies may take part in CSR efforts for a number of reasons, such as their desire to do good deeds, their aim to enhance worker morale, retention, and recruiting, or their desire to provide better services to their clientele. Any reason for an organisation investing in CSR will improve its reputation as a whole. As a result, many businesses view CSR as a chance to enhance their reputation.

The fundamental of CSR is the ethical concept, with changing notions associated with concerns for human welfare and quality of life in the society impacted through the lenses of the dimensions of business activities (Hu, Zhang, & Yan, 2020). CSR can be used as a device to solve community problems as the business works in the society by optimum use of resources (Srivastava, 2017). Businesses are increasingly dedicating significant resources to a variety of social projects, from environmental conservation and community service to ethical corporate practices. In today's socially conscious market climate, this reflects the fact that CSR has a significant position on the corporate agenda around the world.

Owing to CSR's emphasis on social and environmental issues, businesses may use CSR funding to enhance their environmental performance (Prasad, Mishra, & Bapat, 2019). Organisations are under increasing pressure to exhibit exceptional and accountable corporate responsibility as they are considered major builders of a better world from a CSR perspective (Pinkston & Carroll, 1994). According to Ahmad (2016), CSR activities have been influenced by the growth of governance structures and a higher social expectation of corporate responsibility. CSR organisations strengthen their identity and reputation to portray themselves as responsible and good citizens.

The Evolution of CSR

CSR has the hallmark of being a truly global idea. Applicable worldwide, CSR could, in practice, be absorbed by all businesses within the their available resources (Masoud, 2017). According to Gjølberg (2009), CSR has become a fundamental economic, social, and political issue for both developed and developing countries at all levels. Stakeholders have come to trust and respect businesses with a lengthy history of CSR activity (Vanhamme & Grobben, 2009) and extrinsic CSR motives, where the firm is perceived as trying to boost profits while acting with sincere care for the issue at hand (Sheehy, 2015). Looking back at the earlier practices of corporate social contribution, during the 1900s, the corporate contribution was perceived as unfavourable as it was giving away stockholders' money. Still, from 1930 onwards, corporations were seen as institutions with social responsibilities, like the government (Mishra, 2019).

New studies have been prompted by the growing interest in CSR, which sits at the nexus of globalisation, competitiveness, and sustainability concerns. As Asongu (2007) mentioned, organisations' business should be conducted ethically, as a duty to be a good corporate citizen, and in a sustainable manner, ensuring that CSR activities are sustainable. CSR is viewed as the focal point for communication; research suggests that action cannot exist without communication, and that CSR communication improves social action (Christensen et al., 2007). New rules and regulations that drive business activity to address various social concerns have proliferated as a result of globalisation. Political, cultural, and national variables that form CSR practices have an impact on CSR as well (Ahmad, 2016).

CSR has also been a tool to create a win–win situation that helps organisations brand positively (Saxena & Mishra, 2017). The Global Reporting Initiative (GRI) has made CSR a new mandated initiative, while ISO 26000 Social Responsibility addresses the necessity and offers a unified structure (Sheehy, 2015). CSR objectives and purposes have evolved (Rodriguez-Gomez et al., 2020; Nehme & Wee, 2008). The emergence of CSR could be related to social and ecological problems, as businesses have led to significant environmental footprints and social costs for which they have not been held accountable despite their growing financial power. CSR is a concept that has endured and continues to grow in importance and impact, and CSR initiatives could be favourably perceived by stakeholders (De Roeck & Delobbe, 2012).

There has been cynicism around CSR, seen as a smokescreen to hide behind. Blomgren (2011) asserted that the motivation behind CSR activities is more complicated than merely the idea that businesses may be strong agents of social change because of the numerous benefits that could come from such endeavours. In order to find economic possibilities and improve social or environmental performance, strategies are designed around recently established concepts (D'Amato & Roome, 2009). Organisations should emphasise the commercial impact of their CSR operations and how their social initiatives are linked to important business indicators, such as consumer equity, employee retention, corporate governance, and risk management, when describing CSR to this stakeholder group. CSR activities can and should serve societal needs while being prominent in the bottom lines of businesses, according to the objectives of companies and their readiness to embrace a "win–win attitude". Motivation is essential and needs to be in the right direction because, without motivation, there would be resistance to change (Arbussa, Bikfalvi, & Marquès, 2017).

CSR Communication: Management of Crisis

A variety of communication channels and techniques are now required as a result of many organisations' shift in business operations to incorporate CSR. An organisation's internal communication is critical to functioning social, administrative, and technical systems. Well-functioning communication is an essential part of the whole process of CSR implementation in any organisation.

CSR and its communication expenses are often viewed as a long-term investment in building corporate image and reputation. CSR strategies are seen as paramount in reaping benefits through a competitive edge. Various kinds of crisis response strategies can be implemented during different situations or crises, as crises are naturally complex and unpredictable (Zhou & Ki, 2018). Consequently, there are needs for more complex CSR communication methods than ever before, because CSR involvement today requires more advanced, situation-based communication strategies and maintained stakeholder awareness as these have been significant determinants in deciding business reputation (Kim, 2019). During the turbulence of a crisis, companies should take a holistic view when they devise a strategy to respond (Zhou & Ki, 2018). During crisis management, the entire system's efficiency is determined by communication efficiency (Bakos & Dumitras, 2017).

As the public expects organisations to explicitly confirm, demonstrate, and embrace CSR, communicating their CSR initiatives has reaffirmed their efforts to practise CSR, especially during challenging times. The connection of a company's sustainability efforts is reflected through organisation resilience and performance during a crisis with a diverse array of managerial practices (Nazlieva, 2021; Austin & Gregory, 2021). Adopting a dynamic, process-based view provides a holistic understanding of the crisis (Tao & Song, 2020), and good communication, remaining alert and flexible, and being capable of changing strategy are needed to manage changing situations (Morens et al., 2020).

CSR communication could be seen as a radical reconfiguration of an organisation's traditional business models into altruistic, socially responsible business models adopting sustainable development principles (Jahdi & Acikdilli, 2009). Communication must be transparent, as trust is built on transparency, which assures sustainability and stability (Bakos & Dumitrascu, 2017). If stakeholders develop mistrust and believe that businesses' social actions are primarily driven by external factors, CSR communication may suffer. CSR activities should be genuinely motivated by the aim of serving the public and community rather than benefiting the organisation. A genuine and authentic CSR initiative would build a strong bond between the stakeholders as businesses cater to expectations from leading brands during a crisis (He & Harris, 2020).

In Malaysia, a classic case study was when Malaysian Airlines encountered several turbulent events that tarnished its reputation with the twin air disasters crisis involving Flight MH370 (*The Daily Star*, 2014), which is yet to be found, and Flight MH17, which was alleged to have been shot down by the Russian 53rd Anti-Aircraft Missile Brigade (*The Star*, 2022). Malaysian Airlines' crisis reaction was perceived as lacking a convincing communications strategy and demonstrating a lack of compassion for the families of the people who were lost. However, owing to Malaysian Airlines' past CSR initiatives, research indicates there was still a positive relationship between Malaysian Airlines and Malaysian society's perspectives on the brand image, and, according to Selvarajah et al., (2012), brand image influences CSR activities in the Malaysian aviation industry. Factors influencing relationships with customers increase sales and generate customer loyalty, which ultimately affects a company's reputation and its performance regarding CSR activities.

The same conclusion was reached by Ham and Kim (2019), who proposed a framework that combined situational crisis communication theory with the Korean model of persuasion knowledge. Their findings showed significant interactions between CSR intentions, crisis nature, and purchasing intention. Additionally, the relationship between inferences about CSR intents, perceptions of CSR history, and purchase intention has a considerable impact on customer behaviour intentions and brand loyalty (Ham & Kim, 2019).

The Paradigm Shift in Corporate Social Responsibility

CSR today is more about offering an opportunity to create sustainable value shared by society and organisations. It is no longer focusing on the social responsibility of any one organisation but more on how the organisation can naturally fit into being an integral part of society.

CSR reflects the organisation's strategic efforts to create values shared between the organisation and society, often achieved by addressing needs and challenges. A rise in CSR has also caused changes in philanthropic measures. A strategic CSR initiative brings a global idea of philanthropy to a community level, fulfilling the needs and demands of social and economic necessity with a more holistic approach. The organisation's long-term competitive position will be improved by charity through intangible advantages in brand recognition, legitimacy, employee loyalty, and economic performance, although philanthropy may not produce immediate economic and financial returns.

There are opportunities and challenges when looking at CSR in the global environment. CSR is not a complete fix for organisational difficulties, nor will it gain the fullest of the opportunities presented; the most critical factor in making an organisation's brand reputation successful through CSR is making the first move to the strategic approach.

All parties involved, from corporations and governments to NGOs and individual consumers, endorse and encourage CSR. Thus, CSR initiatives can be organised into coherent institutional structures that support a particular goal (Marrewijk, 2003). CSR has evolved throughout the world in the past decade. This evolution is also reflected in governments' active participation in encouraging organisations to engage with the community through social or environmental issues. CSR has the potential to indirectly spur development across a wide range of business activities. Companies should carefully assess the most significant kind of innovation in a given industry before selecting the CSR factors that would best support those advances. Hence, encouraging the innovation-based solution is more integrative in thinking. For instance, educational integration may be more successful, whereas a multidisciplinary capstone course may necessitate going well beyond the common scope of a course (Navarro, 2008).

To address problems that affect all of us, such as poverty, violence, injustice, education, public health, labour, and human rights, McKinsey (2019) proposed the idea of sustainability in four equal components. The social component is one. Providing for basic needs such as food, drink, shelter, and comfort is the responsibility of the economic component. The third component is cultural, protecting and valuing diversity through which communities express their identity and cultivate traditions across generations. The fourth, environmental, component

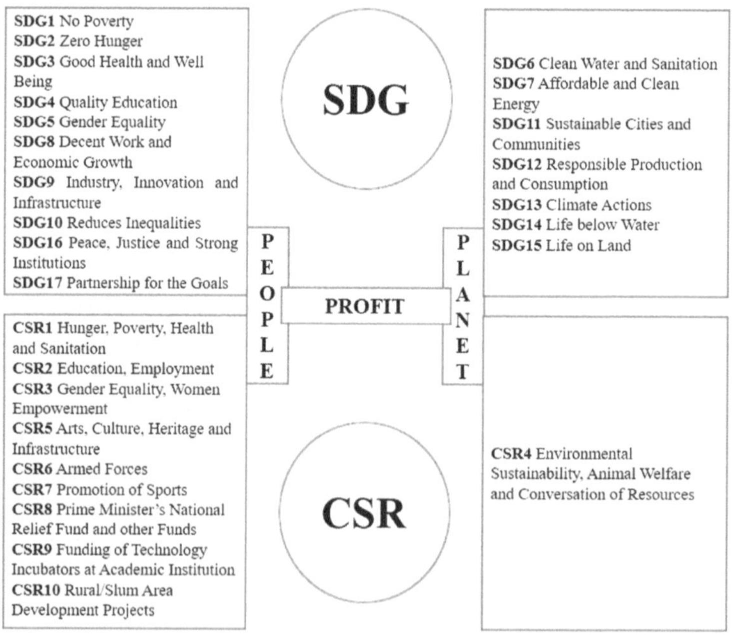

FIGURE 12.4 Overlapping function of CSR and SDG (Mishra, 2021).

works to protect and restore the Earth by preventing waste, preserving natural resources, and controlling climate change.

The Sustainable Development Goals (SDGs), which were adopted by world leaders at the United Nations Conference on 25 September 2015, are the central agenda for sustainable development in Malaysia. The SDGs are a continuation of the development strategy. In Malaysia, the 17 SDG goals are ongoing objectives of the 2030 Agenda for Sustainability, which emphasises the social, economic, and environmental facets of sustainable development. The 2030 Agenda represents a worldwide commitment to more inclusive, resilient, and sustainable growth for Malaysia. Therefore, corporate social responsibility and university social responsibility (USR) will be able to work in tandem with the government pledge to achieve the Masterplan for Green Technology 2017–2030 (Ministry of Energy, Green Technology and Water, 2017).

CSR Education: The Ideal Partnership

A problematic global competitive environment, financial challenges, and the alternative structural and resource commitments to various knowledge areas have increased the challenges learning institutions and organisations face to collaborate effectively. Therefore, many academics are troubled even by the concept of administering higher education institutions in a market-oriented manner, despite the fact that it may be difficult to envisage an alternative to typical businesslike techniques or to revert to the traditional model of the past (Weymans, 2010).

Businesses invest in education for many different reasons, including improved financial performance, reputation building, establishing the organisation as an industry leader, easier access to capital, developing a workforce with more education, raising consumer awareness of a specific issue, and achieving a company mission or mandate. The experience and knowledge that businesses bring to the table can be beneficial to students, educational institutions, and the general public, especially if the organisation and educational institutions collaborate to ensure that the proper demands are satisfied on both sides.

Today's education faces the rising challenges of a practical and comprehensive syllabus, strained budgets, and global competition. Parallel to the growth of the importance of CSR in business, corporation, and social development of the nations, it is time for education institutions to consider the contemporary ideas of CSR in principles and practices. According to Ahmad and Saad (2013), there are four stakeholders in CSR education: the government as policymaker, the community that should benefit from the CSR initiatives, organisations that will be involved and collaborate with education institutions, and the media that need to report on the initiatives. Cross-disciplinary courses taught collaboratively, such as integrative general management, are an example of a multidisciplinary approach. Experience-based activities include everything from simulations of management games and business plan competitions to real-world business and consultancy assignments. Müller-Christ and Liebscher (2015) assert that multidisciplinary integration combined with experiential learning methods would better reflect the real-world business environment, as teams would integrate processes to solve problems, develop new products and strategies, and strategically manage the firm. This would better prepare students with the understanding and grasp of social responsibility for the future.

Previous studies have emphasised the well-established contribution of higher education to the creation of the human capital required for growth. Because of the needs of the knowledge-based economy, government policy priorities centre on the growing role of the higher education sector for nations' socioeconomic advancement (Symaco & Tee, 2019). The competition to recruit top-notch students and academic personnel on a national and international level was another issue that higher education institutions had to take into account in the globalisation era. A framework might offer a means to organise a setting that promotes research for greater comprehension and more impressive synthesis while allowing for the growth of research abilities at all levels and in all fields of study. The current worldwide pandemic has had a big impact on the educational system, as online learning predominates in the learning environment.

Many higher education institutions started utilising a more businesslike approach as a result of the new competitive climate in order to compete and thrive in the expanding education market. The institutions that are not used to thinking and acting strategically are now experiencing ambiguity as a result of this change in academic strategy. In this globalised, post-national world where it is impossible to turn back the hands of time, the only option is to reform colleges by making them more responsive to contemporary economic realities (Weymans, 2010). Ahmad and Saad's (2013) results showing that higher education institutions are participating in USR, which invests in the demands for social well-being and economic, legal, environmental, and charitable duties, were used to support this. This relationship, which is in accordance with national priorities and government policy, will assist organisations in integrating CSR into their strategies.

The research by Ahmad (2012) sheds light on the connection between corporate social responsibility and the development of skills for sustainable lifestyles. It honours cultural diversity and looks for business opportunities in developing the capabilities of workers, the community, and the government. Although CSR has received a lot of attention, little has been said about how university stakeholders, such as students, may help build social responsibility. One may argue that today is the beginning of tomorrow's society, and that, in order for universities to contribute to the creation of this society, their efforts must be motivated, patient, and persistent. It was suggested that education institutions must find ways to meet the demands of the current environment, not only in terms of returns to the community through student projects that include community involvement, but also in terms of replenishing their actions and expanding their sources or references – in other words, to become organisations that contribute to and play a part in creating a new, more moralised society that is active in its neighbourhood and environment.

As education educates tomorrow's workforce, it is no surprise that it has long been a favourite topic, consistently ranked as the most popular, for business community involvement. Teaching business and management at all levels of education should completely incorporate the theory and practice of corporate responsibility and business community participation as essential components of responsible business practices. The abundance of information can be acquired by positioning the higher education sector to be more inclusive and accessible to potential community involvement and capacity-building activities. Here, achievable and realistic cooperation strategies are presented, including a two-way exchange of knowledge and skills between communities and universities.

Higher education institutions now find themselves in a murky environment more akin to a marketplace for consumer goods as a result of recent reforms. This social responsibility principle highlights the more important role that universities play in ensuring and promoting community development that is more sustainable. The majority of colleges emphasise the teaching of CSR using a variety of techniques, with the out-of-classroom approach being the most popular. The difficulties education institutions and organisations confront in collaborating effectively have intensified as a result of a tough global competitive environment, financial issues, and alternative structural and resource commitments to distinct knowledge domains. As it is difficult to imagine an alternative to the conventional businesslike procedures or to return to the traditional model of the past, many academics are bothered even by the idea of governing higher education institutions in a market-oriented manner (Weymans, 2010). Because CSR incorporates larger categories, including human rights, labour rights, environmental stewardship, and social reporting, and is community-driven, CSR practice and instruction should take place concurrently. In order for an educational institution to fulfil its social responsibility commitments, it must take into account the needs of the social context in which it operates as well as the courses and services it provides to society (Ahmad & Saad, 2013).

An Overview of CSR in Malaysia

With a trade to gross domestic product (GDP) ratio that has averaged over 140% since 2010, Malaysia is one of the most open economies in the world (The World Bank, 2018). The expansion of corporate businesses has been fuelled by Malaysia's economic prosperity. To manage a more responsible business, corporate companies are involved in implementing CSR programmes.

Among South East Asian nations, Malaysia is one of the top nations for corporate social responsibility (Wayne, 2008). Owing to the growing number of businesses actively engaging in CSR, the level of awareness and engagement in it is not new. CSR ideas were primarily introduced to Malaysia through the activities of international corporations, claim Amran and Nejati (2014). The goal of CSR

Malaysia is to develop corporate social responsibility awareness and the ability to address social concerns that support responsible business (Ahmad & Saad, 2016).

CSR has gained significance in Malaysia since the country's former prime minister mandated that all publicly traded businesses there include information on their CSR efforts in their annual reports (Abdullah, 2006). The policies and initiatives of the federal government and its agencies motivate organisations to integrate CSR into their business operations (Sharma & Mani, 2013). The introduction, in May 2005, of the Silver Book, a collection of standard CSR principles and criteria for government-linked companies (GLCs), served as yet another major impetus for the practise of CSR in Malaysia (Norhayati & Siti-Nabiha, 2009); see Figure 12.5 for the example of Malaysia National Bank.

CSR initiatives in Malaysia are inclined to have a double impact on extrinsic and intrinsic value (Periyayya et al., 2016). This makes CSR one of the best approaches to establishing an organisation's internal and external relationships. Most organisations incorporate CSR as a means of promotion, showing a particular reputation, or simply being part of a trend. Although some businesses acknowledge the significant importance of CSR, some view it as more of a discrete set of activities or involvement, one-off gestures, or even initiatives that are motivated or driven by marketing, public relations, and business benefits (Amran & Nejati, 2014).

However, most scholars would argue that the critical role of CSR in facilitating long-term and sustainable development, job mobility, and the creation and development of human capital in Malaysia has stirred a lot of research on the many scopes and aspects of CSR enquiries. Similar perspectives were argued at the Asia CSR level, citing that most Asia countries lag in CSR compared with the West, especially concerning the social side of human rights and employment practices (Kim & Moon, 2015). With a few notable exceptions, Asian businesses continue to be more reactive than proactive and lack sufficient stakeholder engagement to define CSR for the business. Few businesses in the area appear to have gone much further than largely separating welfare-related activities from the main company activity. For the majority of them, little evidence of the adoption of CSR practices representing a significant shift in their conceptualisation is seen, which implies an internalisation of certain business processes, whether in supply chain management, human resource management, or labour relations.

Ahmad and Krishnan (2017) assert that the current state of the world necessitates the practice of social and environmental responsibility. Companies can support sustainable development and the preservation of the environment through CSR, both of which can be accomplished ethically through the use of natural resources. The majority of the time, CSR has not been seen as a large, direct source of business revenue, but rather as a burden on business operations, being a link in the supply chain, supplying their goods to bigger organisations or, in rare situations, to the final consumer directly. According to their research, Malaysia's small and medium-sized enterprises (SMEs) are responsible for making sure their products meet the needs of different stakeholders. But, because CSR has

Category: Full Acceptance	Reasons
GLC2	The SB provides guideline towards structured CSR practices, formalized handling of new requests (in dealing with 3rd parties), provides clear CSR direction and business strategy
GLC4	The SB guides to streamlining and formalizing CSR practices, provide guidance in handling new requests, provide guidance in determining how CSR could bring benefits for both parties, the guideline is good and is adjustable to suit the business
GLC5	The SB guides to improve and enhance CSR present system, this guideline can be a government's instruction but it can be adjusted according to nature of business, provide guideline for evaluating new requests, guides towards better planning and execution of CSR, the SB makes things make sense
GLC6	The SB helps in defining area or context of CSR, helps to define on what should be the area of focus for CSR and reduce philanthropic works, provides guidance in entertaining new requests for CSR, provide guidelines for more uniform and systematic way of doing CSR, provides better understanding on the full fledge of CSR activity, the guideline is flexible

Category: Half Acceptance	Reasons: (1) Half Acceptance (2) (2) Positive Side of the Silver Book (SB)
GLC1	(1) Has been implementing CSR for a long period, the guideline is more relevant to new GLCs or GLCs who are just in the initial stage of implementing CSR, the SB is a bit rigid. CBA calculation is difficult (2) The SB guides to structure formality, systematic CSR implementation, the guidelines could be adjusted to suit the nature of business
GLC3	(1) Has been exercising CSR long time ago, already adopting the GRI framework, guidelines are general and certain parts do not cater out type of business and we have different standards, the SB only has meaning locally, CBA calculation is difficult (2) Provides clear context of extent of CSR, guide to more structured CSR practices such as formalize handling of new requests
GLC11	(1) We have been doing CSR for so many years, so when the SB came into picture it just help us to align and formalize our CSR. Our CSR process has been well documented before the SB been introduced, CBA calculation is difficult (2) A set of guideline and principle which help us to align CSR

FIGURE 12.5 Malaysia CSR Standard (BNM Annual Report, 2005).

Country	MALAYSIA	INDONESIA	INDIA	CHINA
	United Nations Global Compact	United Nations Global Compact ISO 26000: Social Responsibility OECD Guidelines: Multinational enterprises	United Nations Global Compact ISO 26000: Social Responsibility OECD Guidelines: Multinational enterprises	
	Global Reporting Initiative (GRI) Sustainability Reporting Guidelines			Global Reporting Initiative (GRI) Sustainability Reporting Guidelines
	Social Accountability International (SAI): SA 8000 Standard		Social Accountability International (SAI): SA 8000 Standard	
		The UN Guiding Principles on Business and Human Rights	The UN Guiding Principles on Business and Human Rights	
Standard and Guideline	Bursa Malaysia Corporate Social Responsibility Framework	The Equator Principles	ILO's tripartite declaration of principles on multinational enterprises and social policy	Generally Accepted Accounting Principle (GAAP)
	Accounting, Auditing and Governance Standards for Islamic Financial Institutions (AAOIFI Standards)	RSPO (Roundtable for Sustainable Palm Oil)	Institute of Social and Ethical Accountability: Account Ability's AA1000 series of standards	GRI 300: Environmental Standard
	The Silver Book		OECD CSR policy tool	
	Shari'ah Governance Framework (SGF)		Global Compact Self-Assessment Tool	
	Islamic Financial Services Board Guidelines		The SROI Network	
			The LBG model	
			National Voluntary Guidelines on Social, Environmental and Economic Responsibilities of Business	

FIGURE 12.6 CSR standards and guidelines (OECD Guidelines, 2015).

been interpreted as philanthropy, with the implication of giving out financial resources typically taken up by large organisations on a periodic basis and serving as a publicity tool, the barriers that prevent manufacturing SMEs from engaging in strategically inclined, environmentally related CSR were due to the level of understanding. With a superficial view that sees corporate social responsibility as a weakness rather than an industrial strength to be developed for business competitiveness, these factors kill the potential of the socially responsible dimension.

Building Bridges: Education and Corporate Social Responsibility

Viewing innovation through the lenses of sustainability, CSR would be the key element in addressing society's social, economic, and environmental complexity. Higher education plays a vital role in educating citizens in creating responsibility for the future. Creating a balance between economic value and the social mission model is essential to ensure the business can be sustained over a long period (Brock & Steiner, 2009).

To develop and cultivate socially responsible learning committed to creating impact, innovative ideas and ability to recognise opportunities and resources for social problems while building sustainable business models should be incorporated into the content and designation of CSR-related curriculum in universities. One of the challenges in teaching CSR is adapting to corporations' business nature while encouraging them to think along new lines of innovations (Pamies et al., 2011). There is frequently faith in improving the curriculum of the university precisely in the school that is offering business-related courses facilities, as such an initiative will result in improvements not only in more aware and conscious business leaders but also in the popularity and status of the university (Popowska, 2016).

The call for education transformation involves not only preparing a new curriculum but also changes in the teaching pedagogies of CSR programmes, courses, and modules (Cetindamar & Hopkins, 2008). Teaching pedagogies such as role-play, case studies, and group discussions on topics help reflect CSR issues (Mishra, 2013). As such, there should be initiatives to design the curriculum so that the core subjects are aligned to the CSR concepts and topics (Mishra, 2013). The critical issues of CSR, such as sustainability, are not systematically being incorporated into the curriculum nowadays. Transmission of knowledge is done in the traditional way, as are the development and accumulation of knowledge, through universities' research. Benn and Dunphy (2009) argued that research methods are best to incorporate into curriculum changes as they build a sense of ownership of outcomes.

Learning and teaching CSR involve various approaches, but the content should also vary. Idea generation, opportunity identification, exploitation, efficient CSR management, funding for CSR activities, stewardship of resources, routes of business establishment, and the associated market are among the key issues discussed in CSR teaching and learning. The entire organisation's scenarios should expose the learning curve of societal needs and problems that enhance knowledge on how to maximise social change and improve social conditions. Approaching organisations and forming collaborations with the senior executives of actual case studies develop value propositions apart from building industry-related relationships.

Conclusion

Creativity is a salient element in dealing with the COVID-19 crisis, as it involves different ways of thinking that allow new ideas to emerge. Practicality and efficiency are essential drivers of agility that contribute to CSR's success, performance, and sustainability. In today's business world, which is full of unpredictability, agility and responsiveness are essential in adapting knowledge to a different situation that arises, as this encourages innovation and learning, especially in CSR implementation.

Innovations provide guidance and structure that encourage responsible practices. For instance, interactions between the business community, labour organisations, civil society, and the public sector could enhance corporate responsibility. Skills in innovation management would be the most important management techniques for organisations to nurture the conception, development, and implementation of creative ideas. CSR could enhance innovations and better performances, as the primary CSR instruments are to create a positive work environment and reduce uncertainties.

To overcome challenges and become competitive and relevant simultaneously, skills, knowledge, and experience are needed to be hand in hand. Learning and enquiries act as an essential system instrument that allows a company to propel, diffuse, and channel knowledge flows about CSR initiatives throughout the company and beyond. Such knowledge allows the company to establish a corporate responsibility management plan. In every field of business activity, innovation helps companies identify social and environmental problems where CSR principles stimulate innovation. Beyond that, education has been claimed as the foundation of the creative problem-solving task.

To encourage an innovation-based solution, more integrative thinking – for example, more successful integration – may take place at the education level, whereas a true multidisciplinary capstone course may require going far beyond the strategy course framework. While CSR practices are not the same across sectors, there are several aspects that must be considered when analysing the relation between CSR and innovation underlining CSR education. Dissimilarity between academic and business environments has led to barriers to cooperation and communication between these two sectors failing to build mutual trust, hindering the undertaking of actual, practical interactions between them. The attitudes and behaviour of business leaders in business education, research, management development programmes, and training through academic institutions also spread the advocacy of new values and ideas. Setó-Pamies and Papaoikonomou (2016) argue that education cannot be seen as a panacea to resolve all ethical dilemmas. Education can be a place for exposure, interaction, and experiences to produce cognitive and affective change in students.

To achieve a more holistic approach and sustainability in CSR education, the curriculum must include the development of communication, negotiation, critical analysis, and change management skills with the core sustainability concepts. The principles, knowledge of sustainability, and research skills development within the CSR subjects and business cases are designed to prepare students to respond effectively as the change agent for sustainability. Universities could conduct research and development activities based on organisations' case studies, a specific problem or event, and the knowledge, technology transfer, and uniqueness of the companies. This aspect would help improve the level of educational innovations, stimulating thinking, proposing alternative solutions, and proposing future solutions and enhancement.

Practical implementation of CSR gives the basis for constructing case studies that encourage better innovative, experiential, agile teaching and learning methods that will improve the quality of CSR learning. With regard to the speed of learning methods, the ability to produce a quicker response in the case of repetitive threats and robust behaviour should be encouraged. Learning could become more collaborative, problem-centred, or self- and peer-directed. Such integration would help the transformation of CSR from a philanthropic-centric to a more sustainable and strategic drive, and the focus of the theoretical explanations should be on the relationship between CSR and innovation.

Organisations should make efforts to embed CSR as part of their innovations and resilience-building efforts as part of their business model. CSR should be integrated into the organisation because employees do not engage in CSR every day, and the company's CSR practices can be seen as theoretical rather than practical. The establishment of CSR vision, diagnosis of CSR problems, development of a plan for CSR initiatives, and communication could be made with a model that would provide a roadmap for managers in other firms to formulate and implement a CSR plan according to the organisation's strategies and mission.

CSR practices, learning methodology, and impact on the sense of guilt and awareness towards community should be incorporated to create corporate citizens who understand their responsibilities towards society and the underprivileged. The high-performance organisations clearly show greater familiarity with the CSR concept, and stakeholders indirectly pressure them to adopt CSR-oriented strategies. In conclusion, socially responsible companies must be able to work effectively with all stakeholders, regardless of their nature or size, as they are a part of the community. These practices represent a commitment by an organisation to behave ethically and contribute to economic development while improving the quality of life of its workforce, family members, and local community. CSR innovation can be a crucial driver and has proven to influence specific channels such as organisations' competitiveness, strategy development, and framework.

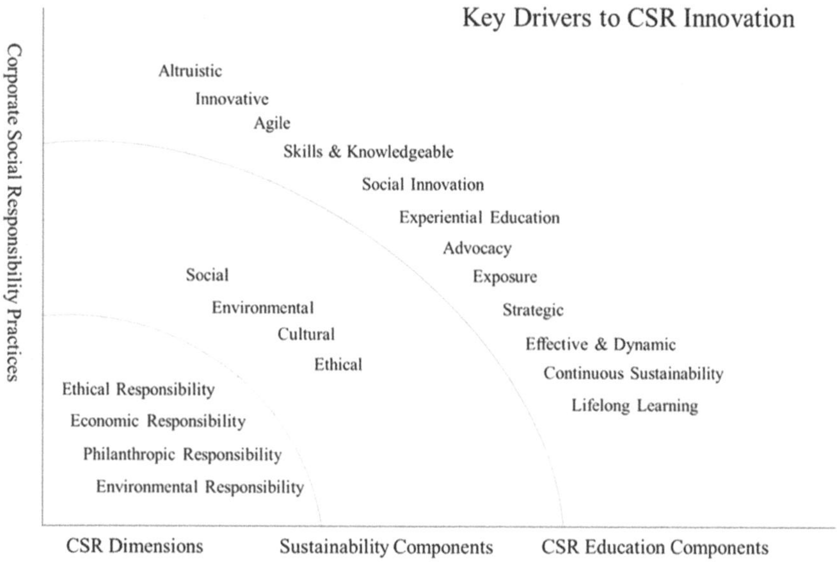

FIGURE 12.7 Key drivers of CSR innovation.

Acknowledgement

The writers would like to extend appreciation to the Ministry of Higher Education Malaysia for the Fundamental Research Grant Scheme with Project Code FRGS/1/2020/SS0/USM/03/8. This chapter is one of the stated tangible outputs. The writers also acknowledge the contributions of individuals who assisted directly or indirectly in completing the chapter.

Bibliography

Aaijaz, N., & Ibrahim, M. D. (2012). CSR inflections: An overview on CSR global strategies adopted by companies in Malaysia and other Asian countries. *Journal of Southeast Asian Research, 2012*, 1–16.

Abdullah, S. N. (2006). Board structure and ownership in Malaysia: The case of distressed listed companies. *Corporate Governance*, *6*(5), 582–594.

Ahmad, J. (2012). Can a university act as a corporate social responsibility (CSR) driver? An analysis. *Social Responsibility Journal*, *8*(1), 77–86.

Ahmad, J. (2016). *Overview of CSR research in Malaysia in 'change, cohesion, competitiveness, Egea publication Italy* (pp. 34–37) Italy: Egea Publication.

Ahmad, J., & Crowther, D. (Eds.). (2013). *Education and corporate social responsibility: International perspectives*. United Kingdom: Emerald Group Publishing.

Ahmad, J., & Krishnan, M. (2017). Environmental related csr initiatives for business strategy: A case study on Small and Medium (SME) Industries in Malaysia. In *The Dynamics of Corporate Social Responsibility* (pp. 305–322). Cham: Springer.

Ahmad, J., & Saad, S. (2013). Beyond theory and practice: A Malaysian case study. In *Education and Corporate Social Responsibility International Perspectives, 4*, (267–296). United Kingdom: Emerald Group Publishing Limited.

Ahmad, J., & Saad, S. (2016). Positioning of corporate social responsibility in media reporting: The role of media setting. In *Corporate Behavior and Sustainability Doing Well by Being Good*. (pp. 121–137).London: Gower.

Amran, A., & Nejati, M. (2014). Corporate social responsibility perception among developing country SMEs: An exploratory study. In *Ethics, Governance and Corporate Crime: Challenges and Consequences*. United Kingdom: Emerald Group Publishing Limited.

Arbussa, A., Bikfalvi, A., & Marquès, P. (2017). Strategic agility-driven business model renewal: The case of an SME. *Management Decision, 55*(2), 271–293.

Asongu, J. J. (2007). Innovation as an argument for corporate social responsibility. *Journal of Business and Public Policy, 1*(3), 1–21.

Austin, Z., & Gregory, P. (2021). Resilience in the time of pandemic: The experience of community pharmacists during COVID-19. *Research in Social and Administrative Pharmacy, 17*(1), 1867–1875.

Axiata Sustainability. (2022). Accessed from https://sustainability.axiata.com/

Bakos, L., & Dumitraşcu, D. D. (2017). Holonic crisis handling model for corporate sustainability. *Sustainability, 9*(12), 2266.

Benn, S., & Dunphy, D. (2009). Action research as an approach to integrating sustainability into MBA programs: An exploratory study. *Journal of Management Education, 33*(3), 276–295.

Blomgren, A. (2011). Does corporate social responsibility influence profit margins? A case study of executive perceptions. *Corporate Social Responsibility and Environmental Management, 18*(5), 263–274.

Brock, D. D., & Steiner, S. (2009). Social entrepreneurship education: Is it achieving the desired aims?. *In United States Association for Small Business and Entrepreneurship. Conference Proceedings (p. 1133). United States Association for Small Business and Entrepreneurship*. Available at SSRN 1344419.

Carroll, A. B. (1979). A three-dimensional conceptual model of corporate performance. *Academy of Management Review, 4*(4), 497–505.

Carroll, A. B. (2008). A history of corporate social responsibility: Concepts and practices. *The Oxford Handbook of Corporate Social Responsibility* (pp. 1–20). Oxford University Press.

Carroll, A. B. (2016). Carroll's pyramid of CSR: Taking another look. *International Journal of Corporate Social Responsibility*, 1(1), 1–8.

Cetindamar, D., & Hopkins, T. (2008). Enhancing students' responsibility towards society through civic involvement projects. *Innovations in Education and Teaching International, 45*(4), 401–410.

Christensen, L. J., Peirce, E., Hartman, L. P., Hoffman, W. M., & Carrier, J. (2007). Ethics, CSR, and sustainability education in the Financial Times top 50 global business schools: Baseline data and future research directions. *Journal of Business Ethics, 73*(4), 347–368.

csrinternational.blogspot.com/2008/11/csr-and-financial-crisis-taking-stock.html. Accessed on 10 June, 2022.

D'Amato, A., & Roome, N. (2009). Toward an integrated model of leadership for corporate responsibility and sustainable development: A process model of corporate responsibility beyond management innovation. *Corporate Governance, 9*(4), 421–434.

De Roeck, K., & Delobbe, N. (2012). Do environmental CSR initiatives serve organizations' legitimacy in the oil industry? Exploring employees' reactions through organizational identification theory. *Journal of Business Ethics, 110*(4), 397–412.

Digi Sustainability. (2022). Accessed from https://sustainability.digi.com.my/

Du, S., Bhattacharya, C. B., & Sen, S. (2010). Maximizing business returns to corporate social responsibility (CSR): The role of CSR communication. *International Journal of Management Reviews, 12*(1), 8–19.

Fassin, Y., Van Rossem, A., & Buelens, M. (2011). Small-business owner-managers' perceptions of business ethics and CSR-related concepts. *Journal of Business Ethics, 98*(3), 425–453.

Friedman, M. (1970). A Friedman doctrine: The social responsibility of business is to increase its profits. *The New York Times Magazine, 13*(1970), 32–33.

Gjølberg, M. (2009). Measuring the immeasurable?: Constructing an index of CSR practices and CSR performance in 20 countries. *Scandinavian Journal of Management, 25*(1), 10–22.

Glavas, A. (2016). Corporate social responsibility and employee engagement: Enabling employees to employ more of their whole selves at work. *Frontiers in Psychology, 31*(7), 796.

Guadamillas-Gómez, F., & Donate-Manzanares, M. J. (2011). Ethics and corporate social responsibility integrated into knowledge management and innovation technology: A case study. *Journal of Management Development, 30*(6), 569–581.

Ham, C. D., & Kim, J. (2019). The role of CSR in crises: Integration of situational crisis communication theory and the persuasion knowledge model. *Journal of Business Ethics, 158*(2), 353–372.

He, H., & Harris, L. (2020). The impact of Covid-19 pandemic on corporate social responsibility and marketing philosophy. *Journal of Business Research, 116*, 176–182.

Hu, B., Zhang, T., & Yan, S. (2020). How corporate social responsibility influences business model innovation: The mediating role of organizational legitimacy. *Sustainability, 12*(7), 2667.

Jahdi, K. S., & Acikdilli, G. (2009). Marketing communications and corporate social responsibility (CSR): Marriage of convenience or shotgun wedding?. *Journal of Business Ethics, 88*(1), 103–113.

Kamal, Y. (2021). Stakeholder expectation disclosure: Evidence from disclosure: Evidence from disclosure: Evidence from disclosure: Evidence from disclosure: Evidence from. *Asian Review of Accounting, 29*(2), 97–127.

Kim, R. C., & Moon, J. (2015). Dynamics of corporate social responsibility in Asia: Knowledge and norms. *Asian Business & Management, 14*(5), 349–382.

Kim, Y. (2019). Strategic CSR Communication in social media: The effectiveness of stakeholder information, response, and involvement communication strategies factoring corporate reputation. *Public Relations Journal, 12*(4), 1–24.

Kulkarni, S. (2017). Inclusion of corporate social responsibility practices as a part of curriculum@ indira school of business studies, india and its impact over the awareness & sense of responsibility of students towards community. *International Journal of Business and General Management (IJBGM), 6*(3), 1–18.

Mahmud, A., Ding, D., & Hasan, M. M. (2021). Corporate social responsibility: Business responses to Coronavirus (COVID-19) pandemic. *SAGE Open, 11*(1), 2158244020988710.

Malaysia, B. N. (2006). *BNM Annual Report 2005*. Kuala Lumpur.

Mark-Herbert, C., Rotter, J., & Pakseresht, A. (2018). *A triple bottom line to ensure corporate responsibility. Timeless Cityland: An interdisciplinary approach to building the sustainable human habitat* (1–7). Uppsala: Sveriges Lantbruks Universitet.

Masoud, N. (2017). How to win the battle of ideas in corporate social responsibility: The International Pyramid Model of CSR. *International Journal of Corporate Social Responsibility, 2*(1), 1–22.

McKinsey (2019). Five ways that ESG creates value. https://www.mckinsey.com/business-functions/strategy-and-corporate-finance/our-insights/five-ways-that-esg-creates-value

Migel, S. P. (2015). CSR reasons, practices and impact to corporate reputation. *Procedia-Social and Behavioral Sciences, 213*, 503–508.

Ministry of Energy, Green Technology and Water Malaysia. (2017). *Green technology masterplan Malaysia 2017–2030.* https://www.pmo.gov.my/wp-content/uploads/2019/07/Green-Technology-Master-Plan-Malaysia-2017-2030.pdf

Mishra, G. (2013). Reflecting Responsible Initiatives for Successful CSR In Context of Higher Education Institutions. *Journal of Business Management & Social Sciences Research, 2*(6), 36–42.

Mishra, L. (2021). Corporate social responsibility and sustainable development goals: A study of Indian companies. *Journal of Public Affairs, 21*(1), e2147.

Mishra, S. (2019). Evolution of corporate social responsibility: Two sets of explanation. *Social Responsibility Journal, 16*(8), 1341–1356.

Morens, D. M., Daszak, P., Markel, H., & Taubenberger, J. K. (2020). Pandemic COVID-19 joins history's pandemic legion. *MBio, 11*(3), e00812–20.

Müller-Christ, G., & Liebscher, A. K. (2015). Transdisciplinary teaching of CSR by systemic constellations. In *New perspectives on corporate social responsibility* (pp. 563–583). Wiesbaden: Springer Gabler.

Musa, A., Yusuf, Y., Mcardle, L., & Banjoko, G. (2013). Corporate social responsibility in Nigeria's oil and gas industry: The perspective of the industry. *International Journal of Process Management and Benchmarking, 3*(2), 101–135.

Nalband, N. A., & Kelabi, S. A. (2014). Redesigning Carroll's CSR pyramid model. *Journal of Advanced Management Science, 2*(3), 236–239.

Navarro, P. (2008). The MBA core curricula of top-ranked US business schools: A study in failure? *Academy of Management Learning & Education*, 7(1), 108–123.

Nazlieva, N. (2021). Corporate social responsibility and employee well-being in light of the COVID-19 pandemic [Master's Thesis, Uppsala Universitet]. https://www.diva-portal.org/smash/get/diva2:1564521/FULLTEXT01.pdf

Nehme, M., & Wee, C. K. G. (2008). Tracing the historical development of corporate social responsibility and corporate social reporting. *James Cook University Law Review, 15*(2008), 129–168.

Norhayati, M. A., & Siti-Nabiha, A. K. (2009). A case study of the performance management system in a Malaysian government linked company. *Journal of Accounting & Organizational Change, 5*(2), 243–276.

OECD Guidelines. (2015). *A comparison of 4 international guidelines for CSR* (p. 7) Danish Standards for Danish Business Authority, *Journal of Business Ethics, 127*(1), 33–47.

Periyayya, T., Nair, G. V., Shariff, R., Roland, Z., & Thanaseelan, D. (2016). Young adult Malaysian consumers' attitude and purchase intentions of CSR supported grocery brands. *Journal of the Southeast Asia Research Centre for Communications and Humanities, 8*(1), 56–77.

PETRONAS Integrated Report. (2020). https://www.petronas.com/integrated-report/files/PETRONAS-IR20-Integrated-Report-2020.pdf

Pinkston, T. S., & Carroll, A. B. (1994). Corporate citizenship perspectives and foreign direct investment in the US. *Journal of Business Ethics, 13*(3), 157–169.

Popowska, M. (2016). CSR education in Poland–current situation and future challenges. *Horyzonty Wychowania, 15*(33), 87–101.

Prasad, M., Mishra, T., & Bapat, V. (2019). Corporate social responsibility and environmental sustainability: Evidence from India using energy intensity as an indicator of environmental sustainability. *IIMB Management Review, 31*(4), 374–384.

Rexhepi, G., Kurtishi, S., & Bexheti, G. (2013). Corporate social responsibility (CSR) and innovation–the drivers of business growth?. *Procedia-Social and Behavioral Sciences, 75*, 532–541.

Rodriguez-Gomez, S., Arco-Castro, M. L., Lopez-Perez, M. V., & Rodríguez-Ariza, L. (2020). Where does CSR come from and where does it go? A review of the state of the art. *Administrative Sciences, 10*(3), 60.

Rosilawati, Y., & Ahmad, J. (2019). Local wisdom and Stakeholder engagement for corporate social responsibility (CSR) initiatives in Indonesia. *Opción: Revista de Ciencias Humanas y Sociales, 21*, 205–220.

Russian suspect in MH-17 airliner downing denies charges on last trial day. (2022, June 10) *The Star.*

Saxena, M., & Mishra, D. K. (2017). CSR perception: A global opportunity in management education. *Industrial and Commercial Training, 49*, 231–244.

Selvarajh, T., Munusamy, J., Chelliah, S., & Sivamurugan, P. (2012). An empirical study on corporate social responsibility in Malaysian aviation industry: A case study. *African Journal of Business Management, 6*(8), 2919–2930.

Setó-Pamies, D., & Papaoikonomou, E. (2016). A multi-level perspective for the integration of ethics, corporate social responsibility and sustainability (ECSRS) in management education. *Journal of Business Ethics, 136*(3), 523–538.

Setó-Pamies, D., Domingo-Vernis, M., & Rabassa-Figueras, N. (2011). Corporate social responsibility in management education: Current status in Spanish universities. *Journal of Management & Organization, 17*(5), 604–620.

Sharma, E., & Mani, M. (2013). Corporate social responsibility: An analysis of Indian commercial banks. *AIMA Journal of Management & Research*, 7(1/4), 0974–497.

Sheehy, B. (2015). Defining CSR: Problems and solutions. *Journal of Business Ethics, 131*(3), 625–648.

Srivastava, A. K. (2017). Alignment: The foundation of effective strategy execution. *International Journal of Productivity and Performance Management, 66*(8), 1043–1063.

Symaco, L. P., & Tee, M. Y. (2019). Social responsibility and engagement in higher education: Case of the ASEAN. *International Journal of Educational Development, 66*, 184–192.

Tao, W., & Song, B. (2020). The interplay between post-crisis response strategy and pre-crisis corporate associations in the context of CSR crises. *Public Relations Review, 46*(2), 101883.

The World Bank. (2018). *The World Bank annual report 2018*. The World Bank.

Twin air disasters threaten Malaysian tourism push. (2014, July 24) *The Daily Star.*

Van Marrewijk, M. (2003). Concepts and definitions of CSR and corporate sustainability: Between agency and communion. *Journal of Business Ethics, 44*(2), 95–105.

Vanhamme, J., & Grobben, B. (2009). "Too good to be true!". The effectiveness of CSR history in countering negative publicity. *Journal of Business Ethics*, *85*(2), 273–283.

Vázquez, J. L., Lanero, A., & Licandro, O. (2013). Corporate social responsibility and higher education: Uruguay university students' perceptions. *Economics & Sociology*, *6*(2), 145.

Wayne, V. (2008). CSR and the financial crisis: Taking stock. Retrieved from https://www.waynevisser.com/wpcontent/uploads/2012/05/blog_financial_crisis_wvisser.pdf

Weymans, W. (2010). Democracy, knowledge and critique: Rethinking European universities beyond tradition and the market. *London Review of Education*, *8*(2), 117–126.

Zhou, Z., & Ki, E. J. (2018). Exploring the role of CSR fit and the length of CSR involvement in routine business and corporate crises settings. *Public Relations Review*, *44*(1), 75–83.

13

EDUTAINMENT IN SUSTAINABILITY

Can Sustainability Communication Also Be Fun?

Mohamad Saifudin Mohamad Saleh

Introduction: Edutainment and Sustainability Communication

Sustainability is often regarded as an uninteresting topic that is difficult to communicate to the audience. Unsurprisingly, many stakeholders, including scientists, policymakers, and educators, struggle to find an effective strategy for communicating sustainability issues such as climate change (Topp, Thai & Hryciw, 2019). Hence, to engage the audience via sustainability communication, a communicator can instil the concept of education-entertainment, or 'edutainment', to ensure more fun and interesting ways of communicating sustainability messages, particularly in relation to the environment, society, and economy. 'Edutainment' combines the terms 'entertainment' and 'education', suggesting a learning atmosphere that engages and motivates learners to grow and develop by involving their emotions (de Fossard, 2008; Sorathia & Servidio, 2012). The term was first introduced by Robert Heyman from the American National Geography Academic Union, which highlighted the idea of education by entertainment (Aksakal, 2015). In that sense, a communicator could alter the process of sustainability communication from a highly critical approach to a lighter one, moving closer to the audience by blending the concept of edutainment in sustainability communication.

Edutainment is often considered a less formal type of education (de Fossard, 2008); hence, it can be conducted in a more comfortable and relaxed setting such as a café, house, park, or gallery (Anikina & Yakimenko, 2015). Accordingly, edutainment for communicating sustainability messages can take place in informal settings, such as at home, when a family watches a film with ecological ideas that encourage fun, enjoyable, yet impactful discussions. This renders sustainability communication, which is commonly considered heavy

DOI: 10.4324/9781003309642-13

and difficult, more relaxed and easier. Simultaneously, it is a great mechanism to overcome the challenge of formal sustainability education in schools or universities that is unable to automatically change people's sustainable behaviour (Sisson, Grisamore & Jang, 2021).

In addition to what is stated above, edutainment in sustainability communication may also be extended in formal situations of schools and universities by having teachers and lecturers communicate sustainability messages in more interactive and entertaining ways. The use of a variety of formal tools such as infographics is encouraged to attract and appeal to students' attention. In line with this, Mindu, Kabuyaya, and Chimbari (2020) conducted a study on the importance of educating school children on schistosomiasis disease in Ndumo, KwaZulu-Natal, South Africa. They discovered that infographics with diagrams and pictures conveying important information helped students to quickly understand the disease and improve their knowledge. Moreover, specialised tools such as infographics could be utilised by educators whenever they want to communicate sustainability issues relating to complex health issues and diseases to assist the audience's comprehension.

It is important to blend the concept of edutainment in sustainability communication to immerse audiences in the concept of sustainability that may be foreign and unfamiliar. For instance, audiences who have never been involved in wildlife animal extinction discourses may not fully understand the issues and concerns. It is even more difficult to communicate these to them descriptively without any aiding tools. Therefore, the use of informative and attractive pictures, as well as exciting animations and videos of wildlife, may help the audience apprehend the content. The power of visual information conveyed via pictures and animated videos will attract the audience's attention and help them visualise the seriousness of the issue. In other words, it will assist the communicator's intention to communicate such complex environmental sustainability concerns to the audience. The combination of edutainment and sustainability communication will expand communication about sustainability beyond simply providing information, to learning and educating about a particular ecological issue.

The Ability to Choose the Right Edutainment Tools

Many edutainment tools such as infographics, films, songs, poems, and videos are readily available and accessible online. Thus, a sustainability communicator needs to be well versed in the tool selection process so that the visuals match the issues. It is important to note that not every tool in edutainment is a perfect fit to communicate different issues in sustainability. One crucial consideration is the audience and the type of suitable edutainment tool. For instance, when a communicator intends to communicate sustainability messages to senior citizens, they may consider simpler tools such as short videos that do not require extensive reading and a long attention span as for an infographic.

Moreover, to commendably engage audiences in sustainability communication, the communicator needs to be alert in selecting appropriate tools based on specific sustainability issues. Some sustainability discourses, such as climate change, are highly scientific and, hence, require sophisticated tools for their communication to audiences. This could be achieved via the use of graphs and statistics presented on an infographic. In contrast, there are more straightforward issues that directly relate to specific communities such as issues of poverty. In this sense, more relatable approaches in the forms of poems, songs, and films could be realised to demonstrate empathy and engage the community.

Another consideration in ensuring the effectiveness of engaging the audience with sustainability communication via edutainment is the availability of the tools. Sometimes, not all tools are readily obtainable for communicating sustainability messages. For example, though various songs, films, dramas, or poems have the ability to impart sustainability messages, they may have limited or no availability in certain places or countries. In Malaysia, songs and films related to environmental sustainability are limited. Hence, ecologically inclined songs and films that are available, such as 'Hijau' (Green) and *Duyung* (Mermaid), are often repeated to convey environmental messages, especially during green events such as the Earth Day celebration.

In the current networked era, social media are excellent edutainment platforms to be utilised for sustainability communication, especially for sharing ecological information with mass audiences. In light of this, Facebook is viewed as a great platform, particularly for students to connect socially, post opinions, and discuss a topic (Isacsson & Gretzel, 2011) such as sustainability. In Malaysia, Facebook is the most used social media platform compared with others such as Instagram, TikTok, and YouTube. Thus, engaging the Malaysian audience on sustainability issues via Facebook and other social media platforms is important.

Videos, images, texts, and other materials on sustainability can be easily shared not only among Malaysians but also with other people worldwide via social media. Wang et al. (2019) concluded that social media, particularly Facebook and Twitter, are useful for sharing edutainment material, such as *East Los High*, a teen drama series that promotes health to larger audiences and connects them with the cast and producers. Some edutainment tools, such as the climate change game EnerCities, can also be played on social media such as Facebook, creating competitions among the players while they obtain information (Ouariachi, Olvera-Lobo & Gutiérrez-Pérez, 2017).

Drama, Television Shows, and Films as a Nexus for Sustainability Communication

Generally, the film industry, set against a variety of cultural promotional channels (Ramírez-Castillo et al., 2021), is heading towards a more sustainable practice. However, the quantity and quality of films, documentaries, television shows, and dramas associated with sustainability, particularly the environment, are still small compared with other themes such as family, love, and fantasy. This is a global situation that occurs not only in Western countries, but also in Asian nations such as Malaysia. For many years, only a handful of local films and dramas in Malaysia portrayed an environmental sustainability theme for audience consumption. This limitation is associated with the belief that environmental sustainability drama or film is heavy, dull, and overall unattractive to the public.

Despite the scarcity of sustainability-themed dramas and films, Howell (2014) revealed that their visual effects, music, pace, storytelling, and sound effects have often been considered crucial and powerful in communicating sustainability issues such as climate change and influencing the emotion and behaviour of the audience. From a Western perspective, Howell argued that films such as *An Inconvenient Truth* and *The Age of Stupid*, as edutainment material, not only introduce the audience to the urgency of climate change but, more importantly, encourage actions to alleviate it. Concurring with Howell, Doyle, Farrell, and Goodman (2017) concluded that *An Inconvenient Truth* provides the audience with data, knowledge, and the threat of climate change to the planet. Undoubtedly, this type of film genre can play a crucial role as an edutainment tool to achieve the Sustainable Development Goals (SDGs), particularly SDG 13: Climate Action.

Another great example of the effectiveness of communicating sustainability messages to the audience via television shows may be drawn from the American comedy *Sex and the City*. In this case, SDG 3: Good Health and Well-Being comes into play when the show successfully raises awareness of breast cancer. According to Gray (2007), the portrayal of Samantha by Kim Cattrall as a strong, authoritative, and self-assured woman, despite being diagnosed with the disease, gives hope to those who suffer from the same illness. The narrative, therefore, not only entertains, but also, to some extent, disseminates information about breast cancer that is often very technical, heavy, and alarming.

In Asia, South Korean films often highlight sustainability themes in their productions. The movie *Haeundae* (Tidal Wave) presents the natural disaster of tsunami by foregrounding the link between the catastrophe and humans as the agents (Elfving-Hwang, 2015). Essentially, the film features a volcanic scientist character who imparts scientific information about the tsunami to the audience throughout the film (Elfving-Hwang, 2015). Like many South Korean movies, *Haeundae* evokes the audiences' emotions by weaving familial and relationship issues against the backdrop of natural disasters, adding drama appeal and differentiating it from a documentary. Consequently, this genre, which sets catastrophe within dramatic screenplays, is a vital edutainment material that provides comprehensive information about tsunamis to the audience. It includes the causes, prevention, and visualisation of the event, which is in fact a rare occurrence in human history. Though this genre is rarely produced and shown in cinemas, the well-crafted resources should be preserved for edutainment purposes for younger generations.

In Malaysia, a film called *Duyung* (Mermaid) has brought a new experience and perspective to the Malaysian audience in relation to environmental-themed films in the country. Through this film, the audience is introduced to a strong message about the importance of cleaning the ocean through the main character, who is portrayed as someone who likes to collect rubbish around the island as he believes that mermaids would appear in a clean ocean. The messages from the film furnish audiences with important environmental awareness and messages in a country that is surrounded by many islands and beaches.

Another great example of an edutainment tool for sustainability communication on poverty issues is the reality television show *Bersamamu* (With You). *Bersamamu* has aired on one of the local television channels since 2005 and is focused on relaying the reality of the everyday life of ill-fated, underprivileged people in Malaysia to the audience (Abdul Wahab, 2010). It invites corporations to carry out their corporate social responsibility (CSR) initiatives and encourages the Malaysian public to donate to the poor. Members of the audience who follow the show are exposed to the reality of unfortunate communities in Malaysia and introduced to economic and societal sustainability themes and discourses. The show is certainly not glamorous and is not emulated by other mainstream media. However, its resolution in highlighting the plight of the underprivileged is unique and demonstrates the awareness of giant broadcasting media, corporations, and the Malaysian public in the communication of the poverty messages.

Despite the limitations, several recent Malaysian dramas have aired on television and online platforms that are associated with sustainability themes related to health education and the COVID-19 pandemic. *Hadiah Dari Tuhan* (Gift from God), aired on local Malaysian television, is one of the examples of a local drama that presents the reality and challenges faced by the society, especially the middle class, during the pandemic (Azmi, 2020). Similarly, another drama, called *Lockdown*, aired on Malaysian local television, also touches on the issues related to the COVID-19 pandemic. The drama was inspired by real stories and aimed to give the audience lessons in life (Kosmo, 2021). As the COVID-19 pandemic commenced, hit the world, and affected people's lives, this drama genre, albeit fictional, provided the audiences with experience from watching the reality of the pandemic that might directly or indirectly relate to them.

In short, from the above discussions, the examples of films, dramas, and television shows are crucial edutainment tools that help a communicator convey sustainability messages related to the environment, society, and economic issues. It is difficult to communicate sustainability messages to the audience without these tools. As mentioned before, embedding this kind of film, drama, or television show in sustainability communication not only makes it more intriguing but also connects the audiences to sustainability messages, especially via visuals that depict the reality of the issues, unlike merely communicating them descriptively using text. Topp, Thai, and Hryciw (2019) concurred that entertaining videos about climate change, for instance, can increase the cognitive engagement of the audience.

Music as a Tool for Sustainability Communication

From another angle, music is an important tool to engage society and disseminate information to it (Benavides et al., 2021), including sustainability messages. For example, the Vietnamese singer Khac Hung has successfully achieved high levels of views of 101 million using music to communicate about the prevention of COVID-19 (Benavides et al., 2021). The significant and powerful effect of music make it an important tool and approach in edutainment, particularly to create joy in learning about sustainability. Therefore, music, especially songs, is another tool to relay sustainability communication besides dramas, TV shows, films, and others.

Like other tools, songs related to sustainability are surprisingly limited in Malaysia compared with other genres, especially love songs. One of the most famous songs that include messages related to sustainability, such as environmental protection and poverty, is 'Hijau' (Green) by Zainal Abidin. Its lyrics disseminate strong messages about the consequences of irresponsible human activities toward the Earth that would later affect future generations. Another green song called 'Pusaka Rimba' (Forest Heritage), sung by the famous Malaysian songstresses Ziana Zain and Dayang Nurfaizah, similarly delivers messages about forest protection and the impact of development on the forest heritage. Besides the lyrics, the melodies of songs can also evoke emotions in the audience, such as sadness, happiness, fear, uneasiness, and others (Kim & Kang, 2022). Thus, the melody of 'Pusaka Rimba', for example, can connect the audience with the emotions of uneasiness and worry about the impact of development on the forest heritage.

Besides the green songs, some songs, such as 'Budak Flat', sung by one of the famous Malaysian rappers, Aman RA, have carried a strong message about the reality of economic and social gaps in the society living in the city (Azni & Wan Mahmud, 2020). From the lyrics of the song, the audience could visualise the problem of living in a flat, such as dirt and disease. Moreover, the lyrics emphasise the lower socioeconomic status of people living in the flats and the important roles of parents to instil identity in kids (Azni & Wan Mahmud, 2020). Unlike the oldies' green songs such as 'Hijau' and 'Pusaka Rimba', 'Budak Flat' has grabbed the attention of the young generations in Malaysia, who favour the rap genre.

Music acts as a connector to forge bonds between individuals, especially when they spend time together listening to live music at festivals or concerts (Kinnunen, Homi & Honkanen, 2020). Thus, a music festival with a sustainability theme, for example, can be organised as a medium to deliver sustainability messages to the people through songs and activities such as donation, cleaning the area, and providing environmentally friendly foods or drinks, so that they can attend it to bond with friends and family and, at the same time, still receive the messages. A good example is Granatos Live, one of the most famous live music festivals in Lithuania, that caught the attention of about

20,000 attendees and emphasised the good practices of sustainable development through the event (Kačerauskas, Streimikiene & Bartkute, 2021). However, live music festival organisers still need to address certain criticisms about the effect of events on waste issues and carbon emissions caused by travel to the event.

Considering the importance of music and songs related to sustainability, radio stations, for instance, can utilise these songs to raise awareness in the public about environmental, social, and economic issues. With this in mind, sustainability-themed songs, especially green songs, should not only be aired to the listeners during a specific season, such as the celebration of Earth Day. They should be aired more frequently so that listeners get the crucial green messages while enjoying the songs. Importantly, more local music and songs in Malaysia should deliver sustainability messages, especially those associated with poverty, health, environment, diseases, and so on. At the same time, the availability of sustainability songs, should be fully utilised by communicators in communicating sustainability messages in any of their communication media. The creativity introduced into sustainability communication by incorporating songs will make it more fun and less dull.

Games as a Tool for Sustainability Communication

Games are often regarded as an effective edutainment tool in teaching and learning sustainable development as they enhance players' knowledge and strategies (Katsaliaki & Mustafee, 2014). It can be argued that players who actively participate and engage in a game will be entertained and educated, especially if it challenges their thinking and engenders emotional responses (Emblen-Perry, 2018). In light of this belief, games are regarded as an edutainment tool linked to sustainability communication that makes transferring fluid sustainability messages easier and fun.

Usually, the game, including serious sustainability games, is made to target a certain group of players. STRASUS, for example, is a serious sustainability game that targets a company via a storyline related to a series of crises and achievement of sustainability events. The players must strategically find a balance between environmental, social, and economic performance while earning a profit for the company (Xie & von Zuydtwyck, 2021). Hence, this kind of game allows the players from the company to enjoy the game and simultaneously obtain information about sustainability.

In addition, game-based learning forms such as board games are also recognised as highly effective tools for environmental (sustainability) communication. They give players an opportunity to interact socially with each other during the gameplay and enable discussion between them (Fjællingsdal & Klöckner, 2020). The sustainability communication that occurs in the social interaction, especially via dialogues between players, allows the more knowledgeable players to share sustainability information such as climate change, poverty, diseases, and so on, depending on the theme of the board game, with other players, particularly those who lack information.

As Malaysia is rich in unique cultures, many traditional games are available among the community, passed down from generation to generation. Some of these traditional games can be good tools for communicating sustainability messages and values to others, especially from the older generations to the younger ones. For instance, a '*Wau*', also known as '*Layang-Layang*' (kite), is traditionally made at home using old newspapers. Similarly, another traditional game, '*Baling Tin*' (can throwing), is played using old cans and a ball. These traditional games can be used as tools in sustainability communication, particularly among family members, teachers, and students at school, by explaining the value of reusing old newspapers or cans instead of discarding them.

Games make sustainability communication less complex, as the playful element makes the complex sustainability issues more accessible to the public (Fuchs et al., 2021). Hence, as an effective sustainability communication tool, games should be utilised more, especially by companies or educational institutions such as schools and universities. However, as the playful and fun element in games can be regarded as less serious and diverted from sustainability approaches (Maltseva, Fieseler & Trittin-Ulbrich, 2019), it is important that communicators such as teachers, lecturers, and others who use games in sustainability communication clearly explain the objectives of playing sustainability games and find a balance between fun and the sustainability aspects. In other words, a sustainable communicator needs to make players realise that sustainable communication can always be conducted in a fun and interesting manner.

Engaging Celebrities in Sustainability Communication

In discussions on sustainability communication, the role of celebrities, especially in conveying messages to their respective followers and fans, is enormous and significant. The appearance and involvement of celebrities in sustainability communication render complex sustainability information more fun and interesting, in line with the concept of edutainment, whereby education about an issue such as sustainability should be done in an entertaining way. The involvement of celebrities makes sustainability information more appealing and attractive. Among sustainable issues, climate change is the most popular topic that has grabbed the attention and advocacy of many celebrities (Park, 2020). This is because it is regarded as the most worrying global issue that has a serious, all-inclusive effect on human health and the entire country, regardless of economic status.

Celebrity involvement and roles in communicating climate change messages started in the mid 2000s via various activities such as giving speeches at world conferences, engaging fans via social media, and producing documentaries, which benefited climate change communication with larger audiences (Doyle, Farrell & Goodman, 2017). In the debate about celebrities and climate change, Boykoff and Goodman (2009) listed six types of climate change celebrities:

(1) celebrity actors, (2) celebrity politicians, (3) celebrity athletes, (4) celebrity business people, (5) celebrity musicians, and (6) celebrity public intellectuals. The six categories of celebrities involved in climate change communication listed by Boykoff and Goodman (2009) redirect the clichéd conventional definition of celebrities from the film and music industries. In other words, those famous and influential individuals from business, sports, academia, and politics can also be influential communicators in communicating and generating audiences' interest in sustainability messages related to climate change and climate action.

In the West, Leonardo DiCaprio is one of the notable examples of high-profile celebrity involvement in climate change advocacy. In 2016, DiCaprio made his Oscar acceptance speech a huge opportunity to deliver his speech on climate change, which reached 34.5 million live viewers (Leas et al., 2016). With his influence and popularity, the climate change messages can easily reach audiences around the globe. Another celebrity actor is Angelina Jolie, who is recognised for her work in advocacy, especially on child and poverty issues (Samman, Auliffe & MacLachlan, 2008).

In Malaysia, celebrity actors such as Maya Karin are also actively involved in engaging Malaysian society with sustainability issues, especially the environment. Her active advocacy work led to her appointment as the ambassador for River of Life Malaysia, advocating the importance of protecting the rivers. Among other issues, she has retweeted a post from the Turtle Conservation Society of Malaysia on the illegal selling of turtle eggs via one of the e-commerce platforms in Malaysia. This incited anger among people, especially her followers, who targeted the e-commerce platform. As a consequence, the e-commerce platform removed the seller's post from its listing.

Recently, the COVID-19 pandemic also became an important sustainability issue, especially regarding health, which has caught the attention of Malaysian celebrities. One of the most famous world-class Malaysian celebrities in badminton history, Lee Chong Wei, advocated that Malaysians get the COVID-19 vaccine to protect their loved ones (Astro Awani, 2021). Chong Wei's message as a celebrity sportsman was crucial, especially in influencing followers to get the vaccine, as some Malaysians still doubted its safety. Without the involvement of celebrities, it is harder for messages related to vaccines or COVID-19 to reach the audience.

From the discussions, it is apparent that celebrities have served as sustainability role models and icons to society, especially young people. Hence, engaging celebrities is a vital edutainment tool to embed in sustainability communication to better deliver sustainability messages to the audience. In Malaysia, although the involvement of celebrities in advocating sustainability issues is limited, their presence in sustainability communication is undeniably crucial, especially in attracting fans to support the case. Thus, it is important for sustainability communicators to include celebrities in sustainability communication activities such as campaigns.

Conclusion

This chapter discusses the aspects and strategies of embedding the edutainment concept, which emphasises fun learning, in sustainability communication. It illuminates how the concept of edutainment, often discussed in education and learning, can also be applied in sustainability communication. Several edutainment tools are available for communicators, particularly in Malaysia, to communicate sustainability messages to audiences, including films, dramas, television shows, music, and games. Celebrity appearances are also crucial in conveying sustainability messages, often regarded as heavy and difficult, especially to the younger generations. In conclusion, a communicator needs to be well versed in using the tools of edutainment, and especially in choosing the right tool for the right sustainability messages and audiences. More importantly, the communicator needs to balance edutainment and sustainability communication so that sustainability communication will not be too entertaining and lose its purpose.

References

Abdul Wahab, J. (2010). Malaysian reality TV: Between myth and reality. *Jurnal Komunikasi, Malaysian Journal of Communication, 26*(2), 17–32.

Aksakal, N. (2015). Theoretical view to the approach of the edutainment. *Procedia-Social and Behavioral Sciences, 186*, 1232–1239.

Anikina, O. V., & Yakimenko, E. V. (2015). Edutainment as a modern technology of education. *Procedia-Social and Behavioral Sciences, 166*, 475–479.

Astro Awani. (2021). *Selepas divaksinasi, Chong Wei ajak orang ramai mendaftar.* Accessed from https://www.astroawani.com/berita-malaysia/selepas-divaksinasi-chong-wei-ajak-orang-ramai-mendaftar-304041

Azmi, N. (2020, December 29). Drama Hadiah Dari Tuhan bawa realiti kehidupan COVID-19. *Berita Harian.* Accessed from https://www.bharian.com.my/hiburan/lain-lain/2020/12/770104/drama-hadiah-dari-tuhan-bawa-realiti-kehidupan-covid-19

Azni, Z., & Wan Mahmud, W. A. (2020). Representasi sosial masyarakat dalam lirik lagu Aman RA. *Jurnal Wacana Sarjana, 4*(3), 1–14.

Benavides, J. A., Caparrós, C., da Silva, R. M., Lembo, T., Tem Dia, P., Hampson, K., & Dos Santos, F. (2021). The power of music to prevent and control emerging infectious diseases. *Frontiers in Medicine, 8*, 1–10.

Boykoff, M. T., & Goodman, M. K. (2009). Conspicuous redemption? Reflections on the promises and perils of the celebritization of climate change. *Geoforum, 40*, 395–406.

de Fossard, E. (2008). *Using edu-tainment for distance education in community work.* India: SAGE Publications India Pvt Ltd.

Doyle, J., Farrell, N., & Goodman, M. K. (2017). Celebrities and climate change. In H. V. Storch. *Oxford research encyclopedia of climate science.* Oxford: Oxford University Press (OUP).

Elfving-Hwang, J. (2015). Disaster narratives in the South Korean cultural imaginary. *Korea Journal, 55*(3), 8–29.

Emblen-Perry, K. (2018). Enhancing student engagement in business sustainability through games. *International Journal of Sustainability in Higher Education, 19*(5), 858–876.

Fjællingsdal, K. S., & Klöckner, C. A. (2020). Green across the board: Board games as tools for dialogue and simplified environmental communication. *Simulation & Gaming, 51*(5), 1–21.

Fuchs A., Pichler-Koban C., Pitman A., Elmenreich W., & Jungmeier M. (2021). Games and gamification: New instruments for communicating sustainability. In F. Weder, L. Krainer, & M. Karmasin. *The sustainability communication reader.* Wiesbaden: Springer VS.

Gray, J. B. (2007). Interpersonal communication and the illness experience in the Sex and the City breast cancer narrative. *Communication Quarterly, 55*(4), 397–414.

Howell, R. A. (2014). Using the transtheoretical model of behavioural change to understand the processes through which climate change films might encourage mitigation action. *International Journal of Sustainable Development, 17*(2), 137–159.

Isacsson, A., & Gretzel, U. (2011). Facebook as an edutainment medium to engage students in sustainability and tourism. *Journal of Hospitality and Tourism Technology, 2*(1), 81–90.

Katsaliaki, K., & Mustafee, N. (2014). Edutainment for sustainable development: A survey of games in the field. *Simulation & Gaming, 46*(6), 1–26.

Kaˇcerauskas, T., Streimikiene, D., & Bartkute, R. (2021). Environmental sustainability of creative economy: Evidence from a Lithuanian case study. *Sustainability, 13,* 1–18.

Kim, J., & Kang, M. Y. (2022). Sustainable success in the music industry: Empirical analysis of music preferences. *Journal of Business Research, 142,* 1068–1076.

Kinnunen, M., Homi, H., & Honkanen, A. (2020). Social sustainability in adolescents' music event attendance. *Sustainability, 12,* 1–17.

Kosmo. (2021). *Lockdown papar garis realiti, duka lara pandemik.* Accessed from https://www.kosmo.com.my/2021/07/26/lockdown-papar-garis-realiti-duka-lara-pandemik/

Leas, E. C., Althouse, B. M., Dredze, M., Obradovich, N., Fowler, J. H., Noar, S. M., Allem, J. P., & Ayers, J. W. (2016). Big data sensors of organic advocacy: The case of Leonardo DiCaprio and climate change. In B. Barnes, & M. Cieply. *Oscar ratings fell, but all eyes are on the academy.* Accessed from http://www.nytimes.com/2016/03/01/business/media/ratings-drop-for-oscar-telecastearly-data-show.html?_r=0

Maltseva, K., Fieseler, C., & Trittin-Ulbrich, H. (2019). The challenges of gamifying CSR communication. *Corporate Communications: An International Journal, 24*(1), 44–62.

Mindu, T., Kabuyaya, M., Chimbari, M. J. (2020). Edutainment and infographics for schistosomiasis health education in Ndumo area, Kwazulu-Natal, South Africa. *Cogent Medicine,* 7(1), 1–18.

Ouariachi, T., Olvera-Lobo, M. D., & Gutiérrez-Pérez, J. (2017). Gaming climate change: assessing online climate change games targeting youth produced in Spanish. *Procedia – Social and Behavioral Sciences, 237,* 1053–1060.

Park, S. (2020). How celebrities' green messages on twitter influence public attitudes and behavioral intentions to mitigate climate change. *Sustainability, 12,* 1–22.

Ramírez-Castillo, N. A., Müller-Pérez, J., Acevedo-Duque, A., Müller-Pérez, S., González-Díaz, R. R., Campos, J. S., & Ovalles-Toledo, L. V. (2021). Sustainable moviegoer intention to attend cinemas based on the Theory of Planned Behavior. *Sustainability, 13,* 1–20.

Samman, E., Auliffe, E. M., & MacLachlan, M. (2008). The role of celebrity in endorsing poverty reduction through international aid. *International Journal of Nonprofit and Voluntary Sector Marketing, 12,* 1–12.

Sisson, A., Grisamore, A., & Jang, J. (2021) Green practices with reusable drinkware at music and sporting events: A hospitality undergraduate student perspective. *Journal of Hospitality & Tourism Education, 33*(1), 1–13.

Sorathia, K., & Servidio, R. (2012). Learning and experience: teaching tangible interaction & edutainment. *Procedia – Social and Behavioral Sciences, 64*, 265–274.

Topp, K., Thai, M., & Hryciw, D. H. (2019). The role of entertainment in engagement with climate change. *Environmental Education Research, 25*(5), 691–700.

Wang, H., Xu, W., Saxton, G. D., & Singhal, A. (2019). Social media fandom for health promotion? Insights from East Los High, a transmedia edutainment initiative. *SEARCH Journal of Media and Communication Research, 11*(1), 1–14.

Xie, N., & von Zuydtwyck, R. H. (2021). *Strategic sustainability by serious gaming: A case study of STRASUS*. Paper presented at the 15th European Conference on Game-Based Learning, Brighton, UK.

14

COMMUNICATING SCIENCE INFORMATION TO THE FISHING COMMUNITY IN THE PHILIPPINES

Joesyl Marie V. de la Cruz-Aranas, Rona Dhel Cabrias Alingasa, and Daryl Lustracion Superio

Introduction: Fisheries and Aquaculture in the Philippines

Fisheries and aquaculture are an important fabric of the economic make-up of the Philippines. In fact, the country makes a significant contribution of over 4 million metric tons of fish products in the world. With this, the Philippines has become known as one of the top fish-producing countries on a global scale. Locally, it contributes greatly to the national economy, which includes supplying food and nutrition to its people and providing a livelihood to over 2 million Filipino fishers (BFAR, 2021a).

Basically, the fisheries and aquaculture industries are economically, culturally, socially, and ecologically important to all Filipinos. However, critical issues affect these industries including resource management, environment, socioeconomics, policy, and, more importantly, research and information dissemination (Green et al., 2003).

The lack of technical data and information on fisheries and aquaculture had been corrected by the establishment of government and regional treaty organisations such as the National Fisheries Research and Development Institute (NFRDI) and Southeast Asian Fisheries and Development Center/Aquaculture Department (SEAFDEC/AQD). Additionally, some academic institutions in the country, such as the University of the Philippines (UP), have contributed data and information on fisheries and aquaculture. These institutions, among many others, have generated data and information that are readily available. Topics range from fishing techniques, marketing, resource management, fisheries laws, rules and regulations, and farming technologies (Camacho, 1999). Simply put, information exists. It is, however, disorganised, written in scientific and technical language, and oftentimes inaccessible. It requires translation into easy-to-comprehend language that is adaptable to current situations and delivered through

DOI: 10.4324/9781003309642-14

preferred media (Luna et al., 2003). Without proper dissemination programmes and campaigns to circulate important information to fishing communities, fishers will continue to be information-poor.

In addition, factors such as poor education, poverty, ignorance of service providers, and the lack of orientation with new technologies affect the information practices of these people. Because of this, there is little possibility for them to satisfy their information needs aside from informal sources, which most of the time end up being inaccurate, untimely, and limited. These inadequacies in knowledge can lead to detrimental effects on the fishers, their livelihood, and the country's food security (Ifejika, 2016; Shuva, 2017). According to Superio et al. (2019), making information available is not enough, as it must be presented in a format that is adaptive to the information-seeking behaviour of the stakeholders.

The key problem when it comes to fishers being information-poor is not unavailability of information but the ineffective communication efforts made by the agencies concerned. According to Herbst and vom Brocke (2012), it is a matter of their ability to adapt to the stakeholders' preferences for locating needed information. Finding relevant information requires the active participation of the user in the information search process, and this usually reveals different information-seeking behaviours.

Information-seeking behaviour is "the totality of human behaviour in relation to sources and channels of information, including both active and passive information seeking and information use" (Potnis, 2015, p. 49). Finding and gathering information have been carried out by humans since the beginning of time. The study of behaviour when seeking information is a newer field in the domain of information science; it reviews the process of storing and retrieving information, especially in this age where myriad sources of information are available (Esew et al., 2014; Al-Suqri & Al-Aufi, 2015).

This chapter aims to promote the importance of profiling how specific groups consume information in order to plan, manage, and create successful, sustainable, and effective information, education, and communication campaigns effectively and efficiently.

Production, Consumption, and Influence of Fisheries and Aquaculture Information among Fishers in the Philippines

Information and data on fisheries and aquaculture are mostly generated through assessment studies and research surveys that provide indicators on the state of fisheries and aquaculture in certain areas. These data sets and information are usually decentralised and are from varied sources. Recorded and existing information is assorted, from the traditional – usually related to fish production and demographics – to complex – research and development, breakthroughs and innovations, and laws and regulations (Carreon & White, 2003; Platon & Israel, 2001).

This information is then published for dissemination, but it is presented in a way that would only be useful for fisheries and aquaculture graduate students, researchers, and government staff, not for the general public or specific stakeholders. According to Lacanilao (1997), fisheries and aquaculture information is abundant in grey literature. It consists of scientific and technical ideas that could affect policy- and decision-making to improve the fisheries and aquaculture sectors. These data and this knowledge are useful for fishers and people at the grass roots when communicated through a mechanism or system that delivers and disseminates information effectively to fishers, and even the public.

It was suggested that increased scientific knowledge might contribute to a reduction in uncertainties and improve the adaptive capacity of poor and small-scale fishers, aquaculture producers, and value chain actors. Information is new knowledge, which leads to a change in the action of people exposed to it. Hence, information gaps and capacity-building requirements must be identified and addressed through networks of research, training, and academic institutions (Dabbadie et al., 2018; Abdullahi et al., 2015).

In attempts to narrow the gap in information dissemination, the Philippine government and concerned agencies have passed legislation and developed information systems and repositories for people seeking information on fisheries and aquaculture.

Freedom of Information Programme in the Philippines

Established in 2016 through an executive order, the Freedom of Information (FOI) Programme aims to provide open access and full disclosure of records, data, and statistics of interests across various government agencies in the Philippines. There are two ways to make FOI requests by standard paper-based requests or by accessing the online portal called Electronic FOI or eFOI. Focusing on the online portal, the beauty of this system is that it is open access, and any Filipino citizen can make a request as long as he or she presents proof of identification. The top government agencies (Table 14.1) concerned with fisheries and aquaculture participate in this information management system.

Three-year data, from 2016 to 2018, showed that 2,582 users (62%) are researchers and students, 780 users (19%) are members of national government agencies, 338 users (8%) are from civil society organisations, 247 users (6%) are from local government, and 210 (5%) are from the media industry. Half (53%) of the users are making requests for the purpose of research, and more than a quarter (27%) are requesting information for school theses and reports. Only a small percentage (10%) of the users are creating requests for personal consumption. In fact, none of the government agencies mentioned in the table were among the top seven agencies with the highest numbers of eFOI requests (PCOO, 2018).

Based on these data and the procedure for accessing information (Figure 14.1), eFOI targets clients with higher educational attainment. Studies have revealed that the majority of the people in fishing communities tend to receive poor to no formal education, which leads to their lack of orientation with new technologies

TABLE 14.1 List of government agencies concerned with fisheries and aquaculture in eFOI

Agency		*Mandates towards Fisheries and Aquaculture*
	Department of Agriculture (DA)	Leads policy-making for agriculture, including livestock, fisheries, and aquaculture
	Bureau of Fisheries and Aquatic Resources (BFAR)	Responsible for the development, improvement, management, and conservation of fisheries and aquatic resources
	Agricultural Training Institute (ATI)	Manages and conducts agricultural and fisheries training and extension
	Bureau of Agricultural Research (BAR)	Leads and coordinates national agriculture and fisheries research and development
	Bureau of Agriculture & Fisheries Products Standard (BAFPS)	Ensures consumer safety and promotes the competitiveness of agriculture and fisheries products
	Philippine Council for Agricultural and Fisheries (PCAF)	Responsible for the attainment of a conducive policy environment towards a globally competitive and sustainable agriculture and fisheries sector
	Philippine Fisheries Development Authority (PFDA)	Provides post-harvest facilities for the development of fisheries and aquaculture products

Source: Presidential Communications Operations Office (PCOO), 2022

(Ifejika, 2016; Shuva, 2017). This greatly affects the information-seeking activities of the community. Information management systems such as eFOI would do very little to satisfy their information needs. If fishers remain information-poor, the state of the country's economy and people's food security are at stake.

Philippine Statistics Agency

Responsible for conducting and consolidating national censuses and surveys, the Philippine Statistics Agency (PSA) has presented valuable fisheries and aquaculture information through databases on OpenStat, an open data platform. It covers a wide array of topics from economy and finance, industry, trade, and services to, specifically, agriculture, forestry, and fisheries (Table 14.2). Information from PSA-OpenStat is usually used for policy-making as it presents data to show

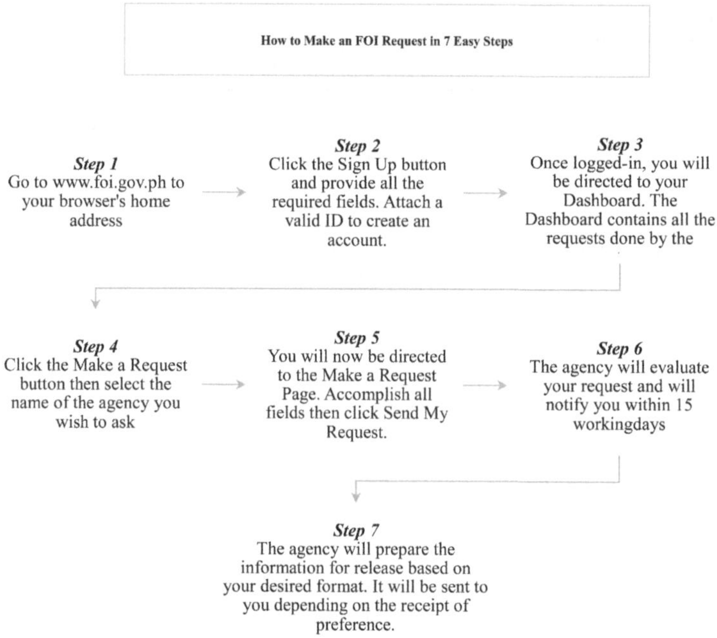

FIGURE 14.1 Procedure for requesting information from eFOI portal (PCOO, 2022).

TABLE 14.2 Databases and information on fisheries and aquaculture contained in the PSA-OpenStat System

Database		*Nature of Information*
Fisheries	Commercial fisheries	Volume and value of production, which can be grouped by geolocation, species, year, and quarter
	Marine municipal fisheries	Volume and value of production, which can be grouped by geolocation, species, year, and quarter
	Inland municipal fisheries	Volume and value of production, which can be grouped by geolocation, species, year, and quarter
Aquaculture		Volume and value of production, which can be group by geolocation, aquafarm type, species, year, and quarter

Source: Philippine Statistics Authority (PSA), 2018

trends in the most in-demand fishery products, volume and value of production, and others. These statistical data became an indispensable and reliable partner for ensuring food security in the country.

Based on its website, the database caters to various clients and stakeholders, but, owing to its open-access nature, there are no records of who accesses the

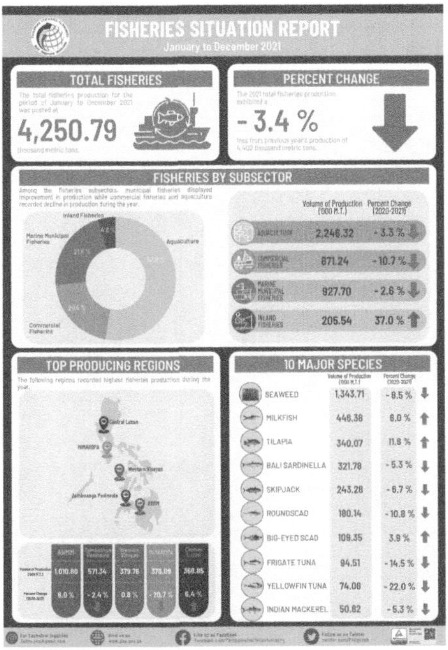

FIGURE 14.2 Sample infographics produced by PSA to supplement data from OpenStat (Source: PSA, 2022).

website. Based on its rather complicated user guide, it can be inferred that the target clientele is not low-income, poorly educated members of the community. However, the agency produces simplified versions of fisheries and aquaculture reports using infographics (Figure 14.2; PSA, 2022).

SEAFDEC/AQD Institutional Repository

This repository was developed as a "reliable means to store, preserve, share research outputs" of SEAFDEC/AQD researchers. But, owing to the need for an effective medium for promoting and disseminating scientific information online and for free, the SEAFDEC/AQD Institutional Repository or SAIR (repository .seafdec.org.ph) has become the official digital repository of the department's scholarly and research information. Launched in 2011, the repository has a collection of over 3,000 scientific papers and publications. The collection also includes in-house publications, fish farmer-friendly materials such as books, handbooks, policy guidebooks, conference proceedings, extension manuals, institutional reports, annual reports, and newsletters. It has recorded over 9 million downloads and 20 million searches since its establishment (Alayon et al., 2013).

There are two ways to access the publications stored in SAIR; one is to directly download open-access materials. Open-access movement encourages

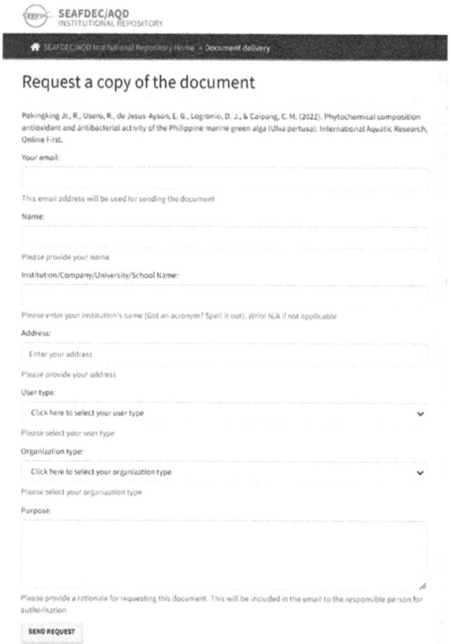

SEAFDEC/AQD
INSTITUTIONAL REPOSITORY

SEAFDEC/AQD Institutional Repository Home » Document delivery

Request a copy of the document

Pakingking Jr., R., Usero, R., de Jesus-Ayson, E. G., Logronio, D. J., & Caipang, C. M. (2022). Phytochemical composition antioxidant and antibacterial activity of the Philippine marine green alga (Ulva pertusa). International Aquatic Research, Online First.

Your email:

This email address will be used for sending the document

Name:

Please provide your name

Institution/Company/University/School Name:

Please enter your institution's name (Got an acronym? Spell it out). Write N/A if not applicable

Address:

Enter your address

Please provide your address

User type:

Click here to select your user type

Please select your user type

Organization type:

Click here to select your organization type

Please select your organization type

Purpose:

Please provide a rationale for requesting this document. This will be included in the email to the responsible person for authorisation

SEND REQUEST

FIGURE 14.3 SAIR request form (SEAFDEC/AQD, 2022).

the free sharing of information from publicly funded research (Alayon et al., 2013). There are over 1,800 items in SAIR that can be downloaded and accessed freely. For non-open-access publications, users have to fill a request form that includes personal information such as a functional email address and the rationale for requesting the document (Figure 14.3). Based on the 2020 report, the majority of the document delivery requests received by SAIR are from undergraduate students, researchers, post-graduate students, business people, and members of the academe. Although many stakeholders have accessed SAIR, as it has collectively served a good number of fish farmers and members of local government units since its inception, the medium and procedure for access may only be suitable for selected members of fishing communities (SEAFDEC/AQD, 2020).

Agricultural Training Institute's e-Learning for Agriculture and Fisheries

The Agriculture Training Institute (ATI) is the official training arm for agricultural extension workers and their clientele. It aims to conduct training courses and provide free print and digital resources to extensionists and stakeholders. The e-learning portal is a website for distance education, and it also supplies technology kits and short video clips on fisheries and aquaculture. Currently, ATI offers techno-kits for milkfish and training courses on how to farm seaweed and cream dory. These are all free when accessed online, but physical copies

TABLE 14.3 Profile of learners using ATI's e-Learning programme from 2009 to 2012 (Solivan, 2015)

Categories	*Percentage (n = 960)*
A. Sex	41.1
Male	58.9
Female	
B. Marital status	38.9
Single	66.1
Married	
C. Educational attainment	4.8
High school	0.3
Vocational	10.4
College undergraduate	56
College graduate	28.4
Post-graduate level	

of techno-kits can be requested for a small fee – PHP100.00 plus shipping fee. To join the courses, you have to enrol through the website during office hours (Monday to Friday, 8 am to 5 pm) as information from the website can only be accessed by registered users (ATI, 2022).

Based on the goals of the portal, its target is to reach both extension workers and farmers to promote rural development. When profiled, however, the users consuming learning materials and training courses from the portal are mostly college graduates, and some even have post-graduate degrees (Table 14.3). Most of them are also well versed in operating typical software applications, and some are even skilled in programming. Although the intention was right, the format of the portal does not resonate with the demographics of fishers in the Philippines (Soliva, 2015).

With all these available scientific data and existing information dissemination efforts, information gaps are still present to this day. Possible answers to enquiries and solutions to the problems of fishers can be found in the data presented in the aforementioned online information management systems. Though free and open to all, these databases are not fully targeted to match the profile of fishers in the Philippines. Their main purposes are government transparency, provision of aid in effective and efficient policy-making, preservation of and open access to research information, and providing information for technical staff and extension workers. These systems do not exist for the sole purpose of reaching small-scale fishers, fish farmers, and other members of fishing communities.

Factors Influencing the Use and Non-use of Fisheries and Aquaculture Information

Todd (2001) described a knowledgeable society as one that is well informed and has the capacity to use information for the betterment of the community.

Information is used in a variety of ways in different sectors. For the fisheries and aquaculture industry, correct and proper use of available information from reliable sources is important to achieve sustainability in catching and farming fish and raising fishers' awareness. Most important is to satisfy the information needs of the fishing community – whether personal or work-related (Durgun, Gunden, & Unal, 2020).

Information-seeking behaviour comprises a set of activities that a person may engage in such as identifying information needs, searching for information, and using and transferring the information gained (Khan & Hedge, 2010). In this process, users are active participants in that they purposefully seek information to increase their knowledge once that information is found. Information-seeking behaviour refers to patterns of actions shown by users to acquire needed information in order to bridge a gap in their knowledge (Abdullahi et al., 2015; Kolarić & Stričević, 2016). Finding and gathering information have been around for a long time, but studying information-seeking behaviour in specific individuals and groups under the domain of information science is what is new. According to Superio et al. (2019), over time, thousands of studies have been conducted on information-seeking behaviour among varying user groups in different industries and almost all aspects of individuals, such as age, gender, economic status, nationality, profession, work, and others. The vast numbers of studies on this subject are due to the dynamism and changeable processes of behaviour in information seeking as it depends on contextual and individual factors. It is vital to understand the psychological processes to understand information-seeking behaviour. Different traits of an individual are associated with the decision to seek information, depending on various psychological factors such as motivation, thinking, perception, memory, interest, cognitive schemes, emotions, and personality traits (Rani, 2017).

The concept of information needs and information-seeking behaviour has been of great concern to stakeholders, particularly minority groups, especially in developing countries where access to information is seen as a mirage (Abdullahi et al., 2015). It happens every day, whether from formal or informal sources. People search the internet, consult experts, ask information from friends, visit libraries, read newspapers, and watch television, among others. Through such activities, people continually form judgements about how useful information is to their particular needs, actively construct meaning, and form conclusions about the relevance of the information to their goal based on various attributes or criteria (Adeyinka et al., 2017).

The earlier part of the chapter presented several formal sources of information on fisheries and aquaculture that are available for access by the public; however, a disconnect between these sources and the fishers still exists.

To explain why this is the case, Dervin's sense-making theory (1983) comes to mind. This theory indicates that there is "a set of assumptions, a theoretic perspective, a methodological approach, a set of research methods, and a practice" designed to cope with information perceived as a human tool designed for making sense of a reality assumed to be both chaotic and orderly. There are four elements to this theory – a situation in which information problems arise; a gap, which identifies the difference between the contextual situation and the desired situation or the outcome as the consequence of the sense-making process; and a bridge – that is, some means of closing the gap between situation and outcome. To bridge this gap, individuals seek information to make new sense and use this information to help them in everyday life. The outcome represents the use of information to complete a task. This makes the sense-making experience a holistic experience situation gap (Kundu, 2017). Dervin presents these elements in a triangle in Figure 14.4.

Sense-making provides a theoretical perspective on information needs, but it is also a methodological approach that could be used to study information usage behaviour (Kundu, 2017).

To integrate the principles, Figure 14.5 shows that the situation refers to the existing knowledge and learned skills of fishers based on their current

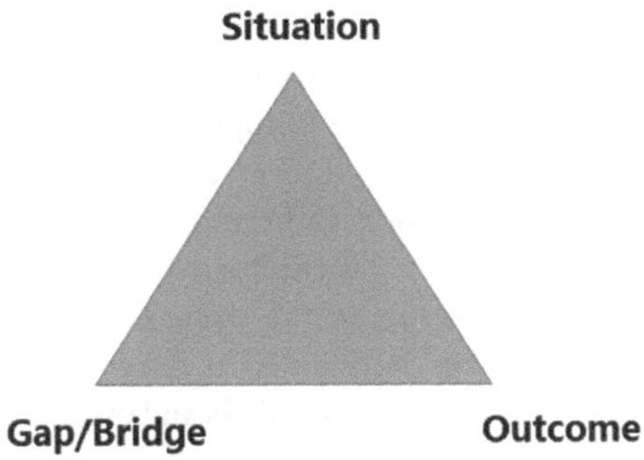

FIGURE 14.4 Dervin's sense-making model (Kundu, 2017)

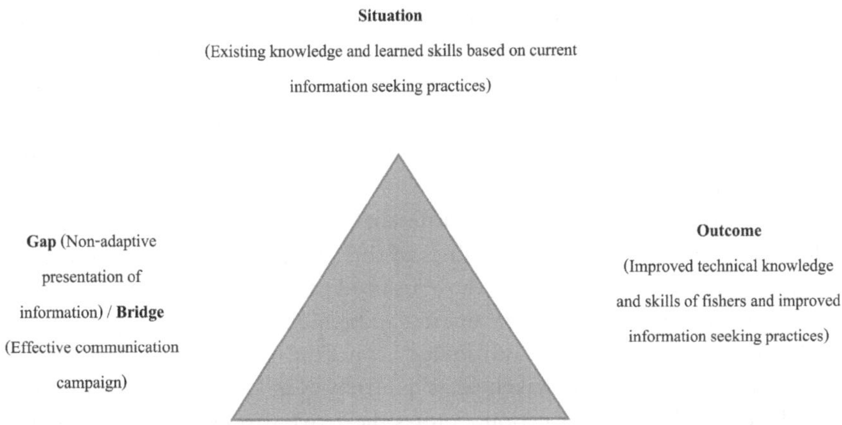

FIGURE 14.5 Integrated framework of information-seeking practices of fishers based on Dervin's model.

information-seeking practices, and the gap refers to the presentation of information that is non-adaptive to the needs of fishers. To bridge this gap, a well-defined communication campaign should be conducted based on the data on the information-seeking fishers' needs. The outcome may refer to the improved technical knowledge and skills of fishers in their livelihood, as well as their improved capacity to seek information from sources built to address their needs.

To this day, fishers are still considered information-poor. Shuva (2017) presented the personal factors that affect the information-seeking activities of fishers, including their low-level education attainment and poor economic status. Most studies reveal that fishers are still mostly illiterate and have a low level of awareness about where to obtain current and relevant information about their livelihood. This shows that there is an acute shortage of information and fisheries extension officers in the field of information dissemination (Njoku, 2004). Accessing formal information sources and services requires extra effort and resources, and fishers tend to have limited orientation to new technologies and are ignorant about service providers (Ifejika, 2016). The detachment between the sources and the stakeholders shows that sectors in fishing communities were not reached by these government and institutional innovations for exchanging information. This indicates that, for countries such as the Philippines, difficulties relating to fisheries information such as inadequate information resources, poor packaging of information, and lack of information-gathering mechanisms has resulted in weak dissemination of information. It was emphasised that there is fisheries information available, but the problems lie in the sharing and dissemination (Superio et al., 2019). Fishers should be considered as fisheries and aquaculture information users, alongside researchers, policymakers and planners, extension specialists, and others.

Adeyinka et al. (2017) also encouraged the provision of many alternatives when it comes to accessing information, instead of being limited to a few unwanted, unfeasible choices. As information continuously plays an important role in the life of humans, the need for information among marginalised groups should be prioritised. A variety of information is much needed for everyone to thrive in this era of information explosion. Everyone needs information to make decisions or achieve respective goals and objectives in life and at work. In other words, there is a need for information in this era, irrespective of person and status, and that includes local fishers.

Fishers are people who perform different types of work and have different roles in a fishing community. Members of this group derive their livelihood in artisanal fisheries and are among those in need of high-quality information to make informed decisions to work smarter and more intelligently (Ifejika, 2016).

As it turns out, information availability was not the main problem. It is a matter of its ability to adapt to the stakeholders' preferences in locating needed information. Finding relevant information requires the active participation of the user in the information search process, and this process usually shows varied information-seeking behaviours (Herbst & vom Brocke, 2012). Fishers, particularly those living in developing countries such as the Philippines, seek and use information in the context of everyday life. Studies have been conducted that described the type of information required by fishers living in developing countries.

Shuva (2017) presented how the fishers in Bangladesh, a country that prides itself on its rivers and is one of the world's leading fish-producing nations, just like the Philippines, practice information seeking. Fishers from communities around the Bay of Bengal require information on the weather, fish buying and selling, entertainment, products, and religion. Likewise, fishers from Nigeria need similar types of information that are essential to allow them to support their fishery livelihood activities, including on the fish markets, finances, fish catch and gear, and weather (Ifejika, 2016). However, information is not sought through organised, systematic information sources. It may not be in a formal setting, but fishers were engaged in their system, such as reliance on personal communication with friends and colleagues, to fulfil their information needs (Uzezi, 2015). In Asia, fishers often share information, on fish prices, markets, fishing and farming techniques, and others, with each other and not through information management systems or any formal sources. According to Superio et al. (2019), fishers felt that their problems and concerns were dealt with and addressed better through personal connections.

There are also barriers present that prevent individuals from seeking and getting information. Information-seeking barriers were defined by Swigon (2011) as a set of physical and immaterial "obstacles hindering, delaying or preventing access to information". According to Shuva (2017), it is evident that fishers have tactically avoided formal information services and providers owing to their unapproachable nature. For example, these providers often impose complicated application procedures and require processing fees prior to giving information – and, more often than not, the information received does not satisfy information needs. Meanwhile, one of the most pressing problems encountered by fishers when seeking information was the harassment received from government officials (Shuva, 2017; Ikoja-Odongo & Ocholla, 2003). The fishing community is considered a vulnerable sector; however, the government has not bothered to create need-based, timely information for it. This calls for a radical change in the information service culture. It is important that various government information agencies offer appropriate, need-based, welcoming information services to rural communities.

These obstacles contribute to fishers being information-poor. The inadequacies in the knowledge of the fishing sector can lead to detrimental effects on the fishers, their livelihood, and the world's food security.

The Case of Fishers in Iloilo Province

Iloilo Province is located 280 air miles from the city capital Metro Manila, specifically in the western part of the Visayas Islands in the Philippines

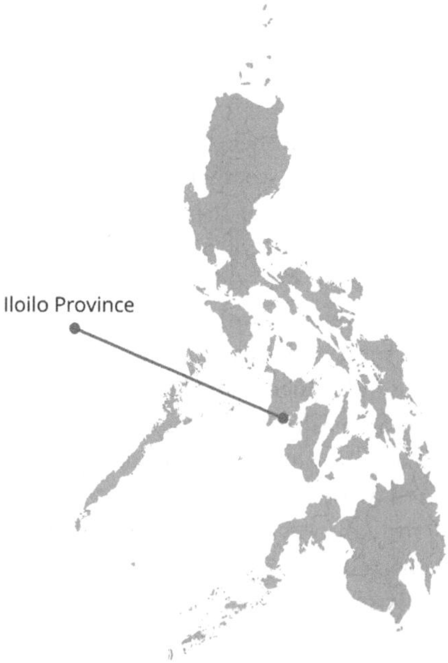

FIGURE 14.6 Location of Iloilo Province on the Philippines map.

(Figure 14.6). It occupies a major south-east portion of the islands, with bodies of water to its east – Guimaras and Iloilo Strait – and Panay Gulf to the south west. The geographical setting of the province allows fisheries ventures and aquaculture production, as well as value-adding business. Iloilo Province belongs to the Western Visayas region, also known as Region 6, alongside six other provinces. Collectively, 200,000 registered fishers in the region are mostly engaged in capture fishing, fish farming, and post-harvest. Among the six provinces, Iloilo Province has the highest concentration of fishers with over 60,000 registered fishers in 2021. In 2021, the province produced a total of 13,454.42 metric tons of fish. There are three fishery subsectors, including commercial and municipal (marine and inland) fisheries and aquaculture (BFAR, 2021b).

Ilonggo (/ɪˈlɒŋɡəʊ/; native of Iloilo Province) fishers are typically older males, aged 41 years and above. This means that fish catching and production are still male-dominated occupations. Furthermore, the majority of them only finished secondary or high school level education, which means that the industry does not require higher education or any formal training (Catedrilla et al., 2010; de la Cruz, 2020; Superio et al., 2019).

Studies, though not yet extensive, were conducted to somehow define the information-seeking practices of fishers in Iloilo Province. Fishers, together with other members of the coastal communities, actively seek information based on their current needs and everyday concerns. According to Felsing, Gonzales, and Pador (2003), fishers in Western Visayas have access to various information sources (Table 14.4). They often seek information on news and current events such as weather updates, personal topics, including health and credit sources, work-related topics such as fishing and culture techniques, and entertainment and leisure (de la Cruz, 2020; Superio et al., 2019).

Fishers frequently seek information about current events in the community and country, specifically relating to the weather. This is information commonly needed by fishers in different parts of the world, particularly in developing countries (Shuva, 2017; Ramadas & Saravan, 2016). The interest in weather and climate reports stems from the knowledge that Iloilo Province and the Philippines as a whole are located in an area that is vulnerable to the effects of the changing environment. It is considered to be one of the natural disaster hotspots of

TABLE 14.4 Media consumed by fishers in Western Visayas

Type of media	*Organization*	*Coverage*	*Language*
Newspapers	*Philippine Daily Inquirer*	National	English
	Philippine Star	National	English
	Manila Times	National	English
	Sun Star	National	English
	Daily Headlines	National	English
	Visayas Daily Star	Regional	English
	The Visayas Examiner	Regional	English
	Bandera	National	Filipino
	Antique Reporter (monthly)	Provincial	Kinaray-a
Newsletters	Candlelight by Jaro, Iloilo City Archdiocese (weekly)	Local	English
	Veritas (weekly)	National	English
	Linkod Tao Kalikasan (monthly)	National	English Filipino
Radio Stations	Agong	Local	Hiligaynon
	DYOK Aksyon Radyo	National	Hiligaynon
	DYKA	Local	Kinaray-a
	Bombo Radyo	National Local	Hiligaynon
	Radyo Natin (community owned)	Local	Filipino Hiligaynon
	Radio Veritas	National	Filipino
	DZRH	National	Filipino
TV Stations	ABS-CBN	National	Filipino English
	GMA	National	Filipino English
	IBC	National	Filipino

Source: Fensing, Gonzales, & Pador, 2003

the world (Combest-Friedman, Christie, & Miles, 2012). Coastal communities in Iloilo Province have been greatly affected by many natural disasters. A few examples are Typhoon Frank in 2008 and Typhoon Yolanda in 2013, which destroyed their homes and livelihoods.

In order to improve their livelihood, work-related information was also actively sought by fishers from Iloilo Province. This includes new fishing and fish farming techniques, marketing strategies and opportunities, post-harvest techniques, among others. Similarly, this specific information need also appeared in studies conducted in other developing countries (Geethalakshmi et al., 2012; Uzezi, 2015).

Aside from livelihood-related information, fishers also seek information for their personal needs, including health and well-being facts, credit and loan sources, products for purchase, among others. Moreover, fishers also actively seek information for leisure and entertainment purposes (Superio et al., 2019; de la Cruz, 2020).

Without proper assessment of their information needs, preferred communication medium, and problems encountered during information seeking, relevant fisheries and aquaculture topics will not be properly communicated and disseminated to them, no matter how important. For example, Superio et al. (2019) evaluated that Ilonggo fishers lack awareness of fisheries laws and regulations, particularly the important stipulations in the Philippine Fisheries Code (Box 14.1). The code directly targets fishers and aims to provide development, management, and conservation of fisheries and aquatic resources. The study concluded that, despite the importance of the code and its provisions, the majority of fishers lack awareness of its existence and objectives. This means that there is a low level of compliance with the rules, regulations, and provisions stipulated in the code, which poses environmental and socioeconomic dangers.

BOX 14.1 PHILIPPINE FISHERIES CODE OF 1998 (REPUBLIC ACT NO. 8550)

Declaration of Policy and Definition

(a) to achieve food security as the overriding consideration in the utilisation, management, development, conservation and protection of fishery resources in order to provide the food needs of the population. A flexible policy towards the attainment of food security shall be adopted in response to changes in demographic trends for fish, emerging trends in the trade of fish and other aquatic products in domestic and international markets, and the law of supply and demand;

(b) to limit access to the fishery and aquatic resources of the Philippines for the exclusive use and enjoyment of Filipino citizens;

(c) to ensure the rational and sustainable development, management and conservation of the fishery and aquatic resources in Philippine waters including the Exclusive Economic Zone (EEZ) and in the adjacent high seas, consistent with the primordial objective of maintaining a sound ecological balance, protecting and enhancing the quality of the environment;
(d) to protect the rights of fisherfolk, especially of the local communities with priority to municipal fisherfolk, in the preferential use of the municipal waters. Such preferential use, shall be based on, but not limited to, Maximum Sustainable Yield (MSY) or Total Allowable Catch (TAC) on the basis of resources and ecological conditions, and shall be consistent with our commitments under international treaties and agreements;
(e) to provide support to the fishery sector, primarily to the municipal fisherfolk, including women and youth sectors, through appropriate technology and research, adequate financial, production, construction of post-harvest facilities, marketing assistance, and other services. The protection of municipal fisherfolk against foreign intrusion shall extend to offshore fishing grounds. Fishworkers shall receive a just share for their labour in the utilisation of marine and fishery resources;
(f) to manage fishery and aquatic resources, in a manner consistent with the concept of an integrated coastal area management in specific natural fishery management areas, appropriately supported by research, technical services and guidance provided by the State; and
(g) to grant the private sector the privilege to utilise fishery resources under the basic concept that the grantee, licensee or permittee thereof shall not only be a privileged beneficiary of the State but also active participant and partner of the Government in the sustainable development, management, conservation and protection of the fishery and aquatic resources of the country.

The state shall ensure the attainment of the following objectives of the fishery sector:

1. Conservation, protection and sustained management of the country's fishery and aquatic resources;
2. Poverty alleviation and the provision of supplementary livelihood among municipal fisherfolk;
3. Improvement of productivity of aquaculture within ecological limits;
4. Optimal utilisation of offshore and deep-sea resources; and
5. Upgrading of post-harvest technology.

Speaking of dangers to the environment, Ilonggo fishers were also surveyed to determine their level of awareness of the effects of climate change (Box 14.2.). It was revealed that local fishers from Iloilo Province were aware of the common effects of climate change but had little understanding or grasp of the long-term effects of the phenomenon (de la Cruz, 2020).

BOX 14.2. EFFECTS OF CLIMATE CHANGE IN FISHERIES AND AQUACULTURE

Ilonggo fishers are extremely aware that the sea level surrounding the province is slowly rising. In fact, a study by a science organisation in New Jersey, USA, found out that the rising seas could wipe out Iloilo City and its neighbouring municipalities.

A few of the more alarming effects of climate change that were noticed by fishers from Iloilo Province include erosion of coastal areas, more frequent occurrences of natural disasters, for example, floods, storm surges, and typhoons, inland waters losing transparency (water turbidity), changes in water colour in oceans, rivers, and ponds, among others.

It can be implied that these levels of awareness were caused by their personal experience as these threats directly affect their livelihood and the communities where they reside.

Information regarding fisheries and aquaculture laws and policies and effects of climate change on aquatic resources are often found in government records, technical reports, scientific journals, and other grey literature. Noting that fishers opt to seek information through personal or face-to-face communication, concerned agencies, such as the Bureau of Fisheries and Aquatic Resources (BFAR), conduct information exchange activities through extension and partnership with fishers' associations to aid this information gap. The BFAR Region 6 office invites experts from research organisations such as SEAFDEC/AQD and academic institutions to conduct training seminars for local government units and community members. Common topics include fishing and fish farming techniques, marketing and economics, resource management, and even laws and policies. Although the seminars are usually a form of personal communication, they are aided by visual materials such as videos or digital presentations in areas with electricity and flip charts and posters for those without. Demonstrations or hands-on exercises are also conducted, particularly on topics such as fishing techniques and fish farming operations. It is also noted that communication initially starts through official channels but moves on to information interactions as relationships between networks progress.

Aside from personal communications, BFAR Region 6 also conducts mass information dissemination through radio shows, where experts give interviews on local radio networks. Information is also disseminated through television and newspapers. Mass information dissemination through the distribution of leaflets, posters, and booklets, mainly written in English and Filipino, is also carried out. Church was also identified as one of the best ways to disseminate information.

However, these efforts have been waning for years. In a report in 2003, the regional office of the BFAR reduced extension activities owing to limited budget and human resources. In addition, the municipal fisheries sector of Iloilo Province received less than 2% of the total municipal budget. These constraints have led to reduced frequency of information campaigns and extension activities and even a lack of human resources (Hablo, 2001; Felsing, Gonzales, & Pador, 2003).

Familiarisation with the stakeholders' information-seeking behaviours can address the aforementioned issues. It can help streamline information dissemination efforts and, hopefully, use only the manpower necessary to get the job done. For example, in the study by Superio et al. (2019), fishers seek information regarding fish farming activities through printed materials such as brochures. Printed publications are suitable for presenting technical and educational information as they can hold multiple pieces of information, can be illustrated, and are easy to reference. Print is also one of the best supplementary materials for other media, particularly personal communications, as it can be taken home for future reading (Ongkiko & Flor, 2006). As fishers are audiences with lower levels of education attainment, publications should be written in Hiligaynon, the local dialect of Iloilo Province, and should have coloured illustrations and photographs (Felsing, Gonzales, & Pador, 2003).

Meanwhile, when it comes to the effects of climate change, particularly information on incoming typhoons and monsoons, fishers prefer listening to the radio for updates. Television comes as the second most preferred medium. This is probably because fishers seek information on a medium that can give them fast, accurate information in real time (de la Cruz, 2020). Mass media channels, such as radio and television, have been the most preferred during natural calamities, disasters, and crises. According to Ongkiko and Flor (2006), radio is easily available in rural communities, requires little to no electricity, and is usually in the local dialect. Community radio stations also exist where they host school-on-the-air programmes on various subjects, including fisheries and aquaculture.

Television, the closest medium to personal communication, was also preferred by fishers, and, as it presents information using sounds and pictures, it can also reach audiences with a low level of education. For fishers from Western Visayas, ABS-CBN is the most preferred news channel as it is free-to-air and has clear reception compared with other TV stations (Ongkiko & Flor, 2006; Felsing, Gonzales, & Pador, 2003). Studies on the impact of the ABS-CBN shutdown have yet to be updated, but reports on Typhoon Ulysses, which caused billions of damages in the fisheries and agriculture sectors in 2020, cited that communities were unprepared for the possible impacts of the storm owing to the lack of an information source, which was television (Marquez, 2020). This was a few months after the non-renewal of ABS-CBN's franchise in June 2020.

Surprisingly, the internet, and specifically Facebook, is an emerging medium of choice for fishers. However, this may not come as much of a surprise as, according to studies, one in every five Filipino adults consumes daily news and information directly from Facebook (SWS, 2019). Recently, television and radio channels have amped up their virtual presence by uploading and streaming shows online through Facebook and YouTube. Aside from mass media platforms, concerned agencies and competent authorities in fisheries and aquacultures, such as BFAR and SEAFDEC/AQD, have garnered huge followings on Facebook (de la Cruz, 2020).

As information seekers, fishers have experienced anxiety while attempting to access information from sources. Fishers have low self-confidence when it comes to seeking information from formal and institution-based sources, including government offices and other fisheries and aquaculture agencies. These inferiorities can be attributed to many things, from their perceived low absorption rates in communicating about technical matters with office workers to unpleasant experiences (e.g., discrimination) in the past. Because of these, fishers no longer seek information from formal sources such as agencies, offices, and authorities, as most described it as a waste of time because they can find information in other sources within their locality and community (Shuva, 2017; Okwu, Yahaya, & Obinne, 2011; Ifejika, 2016). This implies that the majority of fishers are experiencing several levels and dimensions of anxiety, which make them struggle physically, mentally, financially, and emotionally.

Ilonggo fishers, in particular, noted that they had experienced problems in seeking fisheries and aquaculture information. First, most fishers encountered issues initiating an information search as they had no idea where to start. Some who did know where to start were concerned with the cost of an adequate medium to access the information they required. Issues such as lack of information in native dialects and their ingrained anxieties about visiting offices of concerned agencies were also mentioned (de la Cruz, 2020).

Conclusion

When communicating scientific and technical information, there are a plethora of considerations that require the attention of information officers and science communicators of concerned and competent agencies. Aside from planning a timeframe, budget, and consultation with partners and community counterparts, conducting a review of the information needs and seeking behaviours of the fishing community members should be considered. This includes profiling the demographics of fishers, including age, sex, educational attainment, and years of experience. This will help describe the type of audience to be targeted and will set the tone of the campaign message. The existing information sources in the community and the fishers' level of knowledge and awareness of the topics for communication should also be assessed. This will provide a clear jump-off point for the information campaign. Collecting information on the audience's preferred medium, as well as problems encountered in the past, will help build the media formats to be produced. This could help agencies to streamline their efforts to produce materials for selected formats only.

For the case of fishers from Iloilo Province, their reliance on traditional media, such as radio, television, and print media, as the main source of information show that they require relevant, relatable, and easy-to-understand communication materials on topics related to fisheries and aquaculture. Subsequently, information officers and extensionists should not generalise that fishers only opt for traditional media. There is increased consumption and confidence in using the internet, specifically social media such as Facebook. Aside from the production of digital materials to be accessed by fishers, there should be a campaign to protect these new digital consumers from disinformation, as falsities on the internet can be left unchecked. This is why personal communication should always supplement whatever communication campaign fits a certain locality – with a leader of a cooperative or association, a town leader, an elected official, or an assigned fisheries and aquaculture expert, who will be available for consultations.

Moreover, the most important aspect, which every communicator should consider when creating and planning an information campaign for minority user groups, is that each campaign should fit the needs and preferences of the target audience. This way, the information gap can be stitched closed and no longer exist.

References

Abdullahi, A. O., Osahon, I. M., & Esther, S. O. (2015). Assessment of information needs and seeking behaviour of undergraduates in university of Ilorin, Ilorin, Nigeria. *Assessment*, 5(4), 1–16.

Adeyinka, T., Ayeni, J. A., & Oleniyi, O. T. (2017). Assessment of information seeking behavior of physically challenged students in selected Nigerian tertiary institutions. *Journal of Balkan Libraries Union*, 5(2), 24–33.

Al-Suqri, M. N., & Al-Aufi, A. S. (Eds.). (2015). *Information Seeking Behavior and Technology Adoption: Theories and Trends.* Hershey, PA: IGI Global.

Alayon, S. B., Nemiz, E. S., Superio, D. L., de la Peña J. G., & Pacino, L. G. (2013). The development of an institutional repository at the Aquaculture Department of the Southeast Asian Fisheries Development Center, Philippines. *Emerald Insights*, 47(4), 384–398.

ATI (Agricultural Training Institute). (2022, March 24). *Information services.* https://ati2.da.gov.ph/ati-main/content/information-services

BFAR (Bureau of Fisheries and Aquatic Resources). (2021a). *Philippine fisheries profile 2020.* Quezon City, Philippines: Bureau of Fisheries and Aquatic Resources. Download here: Accessed from https://www.bfar.da.gov.ph/publication.

BFAR (Bureau of Fisheries and Aquatic Resources). (2021b). Regional fisheries profile 2020. *Hawas.* Accessed from https://region6.bfar.da.gov.ph/cmsFiles/region6/homepageleft/pdf/13206910-6b34-4c02-90a1-12df579ee836(10-28-2021).pdf

Camacho, A. S. (1999). Major issues and strategies on fisheries. rans. *National Academy of Science and Technology, Philippines, 21*, 120–148.

Catedrilla, L. C., Serofia, G. D., Espectato, L. N., & Jimenez, C. N. (2010). Danyag. *Journal of Humanities and Social Sciences, 15*(2), 81–96.

Carreon III, M. F., & White, A. T. (2003). Information management systems for Philippine fisheries. In G. Silvestre, S. J. Green, A. T. White, N. Armada, C. Z. Luna, A. Cruz-Trinidad, & M. F. Carreon III (Eds.), *In turbulent seas: The status of Philippine marine fisheries* (pp. 192–196). Manila, Philippines: Department of Agriculture, Bureau of Fisheries and Aquatic Resources.

Combest-Friedman, C., Christie, P., & Miles, E. (2012). Household perceptions of coastal hazards and climate change in the Central Philippines. *Journal of Environmental Management, 112*, 137–148.

Dabbadie, L., Aguilar-Manjarrez, J., Beveridge, M. C. M., Bueno, P. B., Ross, L. G., & Soto, D. (2018). Effects of climate change on aquaculture: Drivers, impacts and policies. In M. Barange, T. Bahri, M.C.M. Beveridge, K.L. Cochrane, S. Funge-Smith, & F. Poulain (Eds.), *Impacts of climate change on fisheries and aquaculture* (pp. 449–463). Rome, Italy: Food and Agriculture Organization of the United Nations (FAO).

de la Cruz, J. M. (2020). Information seeking behavior of fisherfolks on climate change in selected municipalities of Iloilo [Master's Thesis, West Visayas State University].

Durgun, D., Günden, C., & Ünal, V. (2020). Information source preferences of small-scale fishers in the Aegean Sea coast of Turkey. *Acta Ichthyologica et Piscatoria, 51*(1), 47–52.

Esew, M., Makarfi, A., Goshie, R. W., & Jimada, A. (2014). An overview of users' information seeking behavior on online resources. *IOSR Journal of Humanities and Social Science, 19*(1), 9–17.

Felsing, M., Gonzales, E. M., & Pador E. (2003). *Information access survey: Western Visayas* (p. 57). Philippines: STREAM/NACA.

Geethalakshmi, V., Jeeva, J. C., Balasubramaniam, S., Parvathy, R., & Nasser, M. (2012). Information and training needs of coastal fisherfolk of Ernakulam District in Kerala. *Journal of Global Communication, 5*(1), 9–15.

Green, S. J., White, A. T., Flores, J. O., Carreon III, M. F., & Sia, A. E. (2003). *Philippine fisheries in crisis: A framework for management.* Coastal Resource Management Project of the Department of Environmental and Natural Resources. Cebu City, Philippines: Development Experience Clearinghouse.

Hablo, R. S. T. (2001). Issues and challenges in the planning and implementation of coastal fisheries projects in the Province of Iloilo [Special Problem, University of the Philippines Visayas].

Herbst, A. & vom Brocke, J. (2012). Information seeking strategies in organizational information architecture. In H. Rahmab, A. Mesquita, I. Ramos, B. Pernici. (Eds.), *Knowledge and Technologies in Innovative Information Systems* (pp. 36–50). New York: Springer-Verlag Berlin Heidelberg.

Ifejika, P. I. (2016). Assessment of fisherfolk information seeking behavior with mobile phone for improve extension and advisory services. *Journal of Agricultural Science, 8*(11), 170–178.

Ikoja-Odongo, R. & Ocholla, D. N. (2003). Information needs and information-seeking behavior of artisan fisher folk of Uganda. *Library & Information Science Research, 25*, 89–105.

Khan, K. M., & Hedge, P. (2010). Information seeking pattern at ICFAI business school library, Bangalore: A user study. *India Journal of Library and Information Science, 4*(2), 107–114.

Kolarić, A., & Stričević, I. (2016). Information seeking behavior for decision making in everyday life: A pilot study on adolescents. *Libellarium, 9*(2), 275–308.

Kundu, D. K. (2017). Models of information seeking behaviour: A comparative study. *International Journal of Library and Information Studies*, 7(4), 393–405.

Lacanilao, F. (1997). Continuing problems with gray literature. *Environmental Biology of Fishes*, *49*, 1–5.

Luna, C. Z., Silvestre G. T., Green, S. J., Carreon III, M. F., & White, A. T. (2003). Profiling the status of Philippine marine fisheries: A general introduction and overview. In G. Silvestre, S. J. Green, A. T. White, N. Armada, C. Z. Luna, A. Cruz-Trinidad, & M. F. Carreon III (Eds.), *In turbulent seas: The status of Philippine marine fisheries* (pp. 3–11). Department of Agriculture, Bureau of Fisheries and Aquatic Resources.

Marquez, C. (2020, November 14). *After ABS-CBN shutdown, lack of Ulysses warning made Cagayan residents suffer.* Inquirer. Accessed from https://newsinfo.inquirer.net/1360703/after-abs-cbn-shutdown-lack-of-ulysses-warning-made-cagayan-residents-suffer-robredo

Njoku, I. F. (2004). The information needs and information-seeking behaviour of fishermen in Lagos State, Nigeria. *The International Information & Library Review*, *36*(4), 297–307.

Okwu, O. J., Yahaya, M. A., & Obinne, C. P. O. (2011). Analysis of artisanal fisher folk information needs and accessibility in Benue State, Nigeria. *Asian Journal of Agricultural Sciences*, *3*(5), 408–413.

Ongkiko, I. V. C. & Flor, A. G. (2006). *Introduction to development communication* (2nd ed.). Los Baños, Laguna, Philippines: UP Open University.

PCOO (Presidential Communications Operations Office). (2018). *2018 FOI Annual Report.* https://www.foi.gov.ph/downloads/2018_FOI_Annual_Report.pdf

Platon, R. R. & Israel D. C. (2001). *The Bangkok Declaration and Strategy for Aquaculture Development Beyond 2000 and Philippine Aquaculture.* Philippine Institute for Development Studies Discussion Paper Series No. 2001-21.

Potnis, D. (2015). *Information seeking behavior and technology adaption: Theories and trends.* Knoxville: IGI Global Publishing.

PSA (Philippine Statistic Agency). (2022). About OpenStat. Accessed from https://openstat.psa.gov.ph/About

Ramadas, G. and Saravanan, P. (2016). Use of electronic gadgets for information seeking process by fishermen in Tamil Nadu: A study. *Journal of Chemical and Pharmaceutical Sciences*, *9*(1), 363–368.

Rani, S. (2017). Psychological factors associated with information seeking behaviour among library users. *International Journal of Library and Information Studies (IJLIS)*, 7(3), 203–205.

SEAFDEC / AQD. (2021). *2020 SEAFDEC/AQD highlights.* Tigbauan, Iloilo, Philippines: Aquaculture Department, Southeast Asian Fisheries Development Center.

Shuva, N. Z. (2017). The information practices of the fishermen in the Bay of Bengal, Bangladesh. *Open Information Science*, *1*, 21–40.

Social Weather Stations. (2019). Fourth Quarter 2019 Social Weather Survey Special Report: 45% of adult Filipinos are Internet users. Retrieved from http://sws.org.ph/

Soliva, M. P. (2015). Determinants of learners' intention to continue using the e-learning program of the Agricultural Training Institute (ATI). *International Journal of Environmental and Rural Development*, *6*(1), 114–119.

Superio, D. L., Nemiz, E. S., Oliveros, M. G. H., Palcullo, V. E. V., Yap-Zerrudo, A. M. A., & Canaman, J. B. (2019). Responsible aquaculturists: The information-seeking behavior of milkfish farmers in Iloilo, Philippines. *Information Development*, *35*(5), 767–780.

Swigon, M. (2011). Information barriers in libraries: Types, typologies and Polish empirical studies. *Library Management*, *32*(6–7), 475–484.

Uzezi, O. P. (2015). Information needs and seeking behavior of migrant fisheries: A case of Isoko Riverine Communities Delta State, Nigeria. *International Journal of Scientific & Technology Research*, *4*(8), 284–288.

CONCLUSION

Sustainability communication is the dissemination of information and interaction between humans pertaining to social, economic, and environmental issues. It aims to make communities more mindful of their actions in certain situations that may directly or indirectly affect the environment, economy, and society.

As reiterated multiple times, sustainability communication has been extensively studied from Western perspectives; accordingly, many discussions and discourses are, perhaps, unbalanced. Sustainability communication as a topic in Asia is still in its infancy, but its importance and impact are undeniable. This book is, therefore, an attempt at representing sustainability communication trends and issues in selected Asian countries with more clarity in an effort to catch up with the Western hemisphere.

The articles in this book are presented as chapters that cover a wide range of themes, including language, economy, content marketing, gamification, journalism, community radio broadcasting, corporate social responsibility, edutainment, as well as fisheries and aquaculture. These topics are presented and discussed in the context of the practice of specific communities in Malaysia, Indonesia, Thailand, the Philippines, and China. In some chapters, case studies further substantiate the sustainability communication endeavours adopted by the populations. Before concluding, let us review the key takeaways from each chapter.

Chapter 1 of the book begins with a debate on the similarities and differences in sustainability communication through the Western and Asian lenses. The chapter then moves on to understanding the integrative approach of sustainability communication that encompasses environmental, climate change, and science communication. Essentially, readers learn that sustainability communication progresses alongside technology, whereby messages are now disseminated almost exclusively online rather than on traditional platforms. Communication

DOI: 10.4324/9781003309642-15

technology was never more prevalent than when the world was hit by the COVID-19 pandemic, which saw a rise in the utilisation of digital media to connect, disperse, and promote sustainability agendas at local and global scales. The chapter ends by highlighting the scarcity of studies concerning sustainability communication in the Asian context. This, in fact, is an opportunity for many Asian researchers and scholars to venture into this field and provide as much literature as possible to match the Western hemisphere. The uniquely distinct Asian culture, economic background, and locality that play pivotal roles in sustainability communication may contribute to different outcomes compared with the other side of the world. Henceforth, the chapter encourages more research on sustainability communication to be conducted using an Asian lens.

In Chapter 2, readers discover the language of sustainability communication and how language impacts the portrayal and image of an organisation. It starts by providing a brief background of Malaysia's sustainability issues and communication, before introducing the language of sustainability communication, which proposes illocutionary acts as the basis of sustainability writing. The chapter deliberates on the issues of content, linguistics, and visual analyses in sustainability communication, as these form the language foundation for message creation. In demonstrating sustainability language in action, a case study of Universiti Sains Malaysia's (USM) Facebook posts is presented. It concludes that USM shows strong advocacy for economic and societal sustainability agendas, which is in line with its function as a higher education institution. The chapter ends by highlighting the value of compelling "SusTelling", or sustainability storytelling – conceptualised with the right lexicogrammar, personal pronouns, inclusive tones, and active constructions – to forge a robust connection with the audience and inculcate a sense of belonging to the organisation and future sustainable initiatives.

Chapter 3 invites readers to explore the critical role of the sharing economy and how it boosts China's economic transformation. In particular, it introduces readers to the way the sharing economy in China provides platforms for idle resources and promotes recycling as an agenda to realise sustainable development. The decentralisation of the sharing economy further leads to the incremental growth of user-generated content and spread of sustainability communication. The sharing economy can indeed nurture interest in the public to seek further sustainable information and motivate acts and intention, in line with the objective of a sustainability message. Consequently, this practice leads to effective and strategic two-way sustainability communication between the organisation and the public.

In Chapter 4, readers discover strategic sustainability communication from the economic point of view of companies listed on the Thailand Stock Exchange. It starts with a background of sustainability from Thailand's perspective and the private sector's. It then discusses the process of sustainability communication of private organisations listed on the Stock Exchange. The three main dimensions of strategic sustainability management are the economy, society, and the environment. Two additional dimensions that are also sometimes adopted are corporate governance and risk management. This chapter encourages listed companies to implement strategic sustainability communication by considering and serving stakeholder's expectations and underlines the prominence of digital media in current and future messages. All in all, this chapter would very much benefit corporate communication experts in government and private organisations, ensuring more effective and impactful economic sustainability messages for their stakeholders.

Chapter 5 discusses the impact of new media and persuasion techniques on ecological communication by zooming in on the Penang Green Council's (PGC) initiatives. Drawing upon Aristotle's framework of appeals, the authors analyse the logos/logic, ethos/credibility, and pathos/emotion in PGC's website discourse. Logos is prominently utilised in the messages, especially in the dissemination of factual information, which generates a better understanding of the green efforts. Ethos is seen in PGC's inclination towards instrumental dependency, while pathos is seen in the corpus being used to trigger guilt to further persuade audiences towards better sustainability actions. The chapter ends by summarising PGC's use of personalised, transparent, and motivating messages to move audiences who are already passionate about environmental issues; it also suggests the incorporation of societal and cultural values and norms in its messages to spark interest among those who are new to the field of sustainability and those who lack knowledge about sustainable living.

In Chapter 6, the idea of sustainability communication via social media is put forward. It leads readers to understand the changing world after the COVID-19 pandemic, which inflated demands for sustainable consumption of products and services. Owing to this, organisations are jumping on to the bandwagon of portraying sustainable efforts in their brand communication to garner more returns on investment (ROI) for their businesses. The trend in doing so is to ideate the right and strategic sustainable content in their sustainability advertising. In realising this, the author explains how to stay away from producing greenwashing messages when communicating sustainability intentions. Finally, the chapter deliberates on utilising the right message appeals in sustainability content and the core elements of effective content ideation, particularly for social media marketing.

The practices of young Malaysian esports and gaming influencers and their involvement in sustainable development are elaborated in Chapter 7. Instead of only focusing on playing video games, gamers today also watch live gaming content to acquire more knowledge and forge connections with their esports idols. This leads to a greater demand for live gaming content, where gaming influencers are able to reach out to young audiences. The live-stream sessions of prominent gaming influencers often achieve high engagement rates. Organisations who see this as an opportunity collaborate with these gaming influencers to build relationships and persuade audiences to participate in green initiatives and society well-being, such as fundraising and other community activities.

Chapter 8 invites readers to learn about sustainable development and sustainability communication through the use of virtual exhibitions. This chapter, particularly, explores the Indonesia Geopark virtual exhibition that was conducted in conjunction with the Indonesia Geopark Summit of 2021. The virtual event utilised digital space to ignite public interests in geoparks, as well as removing any barriers to educational activities. Virtual exhibitions via digital spaces offer a remarkable opportunity to promote a sustainable learning environment and tourism activity, especially owing to the enhanced sensory experience afforded by the audio-visual features.

Chapter 9 of the book underlines the need for a sustainable mobile application in China. Specifically, the chapter focuses on gamification and sustainable behaviour among the Chinese users of an app called *Ant Forest*, a game module developed by Alipay that has attracted a large number of users owing to its meaningful engagements and diverse rewards. The gameplay was educational for more than 500 million users, and 100 million of them were persuaded to transform their gameplay into practical sustainability behaviour, where they planted real trees in the desert areas across China to protect the environment. This demonstrates that effective sustainable communication may be initiated via diverse digital platforms and is highly impactful in influencing consumers' actions.

The four factors that encourage journalistic sustainability communication in news writing are elaborated in Chapter 10. Readers are invited to explore the factors of exposure, passion, knowledge, and network that guide a journalist's news writing endeavours. Exposure acknowledges a journalist's experience in conducting sustainability communication journalism. Exposure ignites passion to search for knowledge and create networks. The combination of passion and knowledge raises journalistic capacity, while a network allows wider access to environmental experts and data resources. A journalist should, therefore, be able to identify and associate with relevant bodies and organisations to sustain his/her passion for the job.

Chapter 11 focuses on utilising community radio broadcasting as a medium to increase the capacity of local communities to handle climate change issues. The chapter revolves around Action for Climate Empowerment (ACE) – a framework for climate change capacity building. The framework highlights two elements, which are public awareness and public access information. Capacity building is crucial in increasing the knowledge and understanding of personnel managing community radio operations on climate change. Future directions of the effort should also integrate collaboration between relevant stakeholders such as government agencies, NGOs, companies, and media organisations to transform community radio into an effective medium that raises awareness and spreads information pertaining to climate change.

In Chapter 12, readers get to explore corporate social responsibility (CSR) and the changes that have taken place over the years owing to globalisation, innovation, and education from the point of view of the social dimension and social impact. Future directions of sustainable CSR must be realised via more holistic education empowerment, which includes the topics of communication, negotiation, critical analysis, and change management skills within the core sustainability concepts in its curriculum. Added to this, more efforts to incorporate CSR as part of an organisation's innovation and resilience-building initiative in its business model are highly encouraged.

Edutainment in sustainability is the topic of discussion in Chapter 13. It focuses on how sustainability communication could also contain a fun element and appeal to the audience's consciousness. In educating via sustainability communication, the concept of edutainment allows for a more stimulating experience and a less technical method. Several edutainment tools that are deemed suitable for disseminating sustainability messages include films, dramas, television shows, music, and games, which may also include celebrity appearances. The author reiterates that the choice of the right edutainment tools and relevant celebrities will increase the impact of sustainability messages on the audience.

Chapter 14 focuses on the issue of communicating science information to the fishers in the coastal communities of Iloilo, Philippines. In matching their need for aquaculture knowledge and information-seeking behaviour, readers are exposed to the elements and approaches to consider when crafting scientific and technical guidance. Prior to the distribution of any communication materials, information and science correspondents must conduct profiling of their audience. Profiling ensures the distinctive demographic backgrounds, needs, and preferences are considered when crafting and setting the tone of campaign messages to get the messages across effectively. Moving forward, the authors believe that the media used to communicate with fishers should not be restricted only to traditional ones but should be expanded to include social media as well.

All in all, this book has explored various themes of sustainability communication across Asia. It is interesting to note that sustainability discourses are wide and varied and cover various aspects of life that are subsumed under the triad of environment, economy, and society. The book has shown that sustainability communication is in fact a growing field in Asia and can be further developed and expanded to become a major discipline in this region. More efforts should be undertaken by scholars to understand and represent sustainability communication among Asians.

Notwithstanding, sustainability communication's theoretical and academic contributions should be matched with its practical, real-world applications, especially when navigating a world with depleting natural resources and high consumption patterns. In other words, the sustainability communication effort should have a transformative impact on members of the public to persuade them to become more sustainable, ecological, and responsible global citizens. One way to do this is by crafting impactful sustainability messages for specific communities, distributed via social media platforms. Social media are indeed connecting, trendy platforms across Asia, used by millions of subscribers. By appealing to their senses, based on their needs and preferences, more sustainability messages may get through and influence actions.

Therefore, it is hoped that this book will contribute by providing greater knowledge pertaining to sustainability communication, from the perspective of the Asian context, for all stakeholders – local as well as global citizens.

INDEX

Note: Page numbers in *italics* indicate figures, **bold** indicate tables in the text.

For Product Safety Concerns and Information please contact our EU representative GPSR@taylorandfrancis.com Taylor & Francis Verlag GmbH, Kaufingerstraße 24, 80331 München, Germany